SMART ENVIRONMENT MANAGEMENT

THE EDITORS

Dr. Neerav Sharma is the Research and Liaison Officer at the Indian Institute of Technology Roorkee with post-doctorate from Purdue University, USA and doctorate from IIT Roorkee in the fields of GeoAI, IoT and transportation systems. With research experience of 6 years, he encompasses a wide variety of research interests ranging from AI, IoT, sensors, UAVs to remote sensing, geospatial and GeoAI applications. With multiple reputed Journal and Conference publications and best-paper awardee at IEEE-ICEET Conference, Malaysia, he also holds memberships at IEEE, ISRS, ISG and acts as reviewers and editorial board members for reputed journals.

Dr. Sandhya S. Lanjewar is the Head of the Department of Botany at Government College, Kohka-Neora, Tilda, Raipur, Chhattisgarh, India. With over 14 years of dedicated research experience, her work primarily focuses on Microbiology. She has an impressive academic portfolio, having participated in more than 37 national and international seminars and conferences. Dr. Lanjewar has published over 25 research papers in reputed journals and has attended 6 workshops. Additionally, she has completed 3 short-term courses and published 2 books. She is an active member of numerous academic bodies, demonstrating her commitment to the advancement of science and research.

Dr. Kavita Sharma currently serves as the Head of the Department of Botany at Government Arts and Commerce Girls College, Raipur, Chhattisgarh. With an illustrious career spanning over 38 years in teaching and 24 years in research, her expertise lies in Mycology, Environmental Biology, Biodiversity Conservation and Remote Sensing-GIS applications. She has mentored more than 20 PhD scholars and has authored more than 160 research papers in renowned national and international journals. Dr. Sharma also holds editorial and reviewer roles for several distinguished journals across the globe. In addition to her prolific publication record, Dr. Sharma has authored 15 books, primarily focusing on Mycology and Environmental Science. She has successfully completed five major research projects funded by the University Grants Commission (UGC) in New Delhi and Chhattisgarh Council of Science and Technology (CCOST) in Raipur. Her academic contributions extend beyond national borders, having visited over 10 countries to present her research findings and delivering expert lectures, often as a keynote speaker or chairperson. Dr. Sharma's extensive body of work reflects her dedication to advancing environmental research and education.

SMART ENVIRONMENT MANAGEMENT

– Editors –

Dr. Neerav Sharma

Dr. Sandhya S. Lanjewar

Dr. Kavita Sharma

ASTRAL

Publishing Authors Across the Globe

2025

Daya Publishing House®

A Division of

Astral International Pvt. Ltd.

New Delhi – 110 002

ISBN: 978-93-5919-213-0

Published by : **Daya Publishing House®**
A Division of
Astral International Pvt. Ltd.
– ISO 9001:2015 Certified Company –
4736/23, Ansari Road, Ground Floor
Darya Ganj, New Delhi-110 002
Ph. 011-43549197, 8130496929
e-mail: info@astralint.com
Website: www.astralint.com

Printed at : Replika Press Pvt Ltd

Dedicated To

LATE SHRI. K. R. SONATKE

पुण्य स्मरण, परम श्रद्धेय पूज्य बाबूजी दिवंगत श्री के. आर. सोनटके जी को सादर समर्पित जिनकी प्रेरणा और मार्गदर्शन ज्ञान की ज्योत बनकर सदैव मेरा उत्साहवर्धन करते रहे।

संध्या

PREFACE

In an era marked by rapid urbanization, climate change, and environmental degradation, the need for innovative and sustainable approaches to environmental management has never been more urgent. The fusion of science, technology, and traditional knowledge offers a promising path toward creating resilient ecosystems and communities. This book, Smart Environmental Management: The role of Science and Society, explores the multifaceted dimensions of sustainability, emphasizing the harmonious integration of cutting-edge innovations with time-tested practices.

The chapters presented here delve into the complexities of climate adaptation, highlighting strategies that not only mitigate environmental risks but also enhance societal resilience. Remote sensing technology, with its powerful applications in monitoring and managing natural resources, serves as a cornerstone for informed decision-making. This book also explores sustainable resource management, bringing together technological advancements and traditional wisdom to ensure balanced ecosystem services. Electric vehicles are another critical focus, representing a paradigm shift in environmental planning and urban development. Through a comprehensive lens, this book considers the role of electric mobility in reducing carbon footprints while examining the larger framework of environmental planning required to integrate this transition.

By weaving together perspectives from science, technology, and traditional knowledge systems, this work aims to offer insights into building a future where environmental, societal, and economic imperatives converge in sustainable harmony. The journey through these topics will provide readers—whether scholars, policymakers, or environmental practitioners—with a roadmap to contribute to a more sustainable and resilient world.

This book is not just a collection of knowledge but a call to action for future innovators, planners, and stakeholders to rethink environmental management through an interdisciplinary approach that values both modern advancements and the deep wisdom of traditional practices.

Dr. Neerav Sharma

Dr. Sandhya S. Lanjewar

Dr. Kavita Sharma

CONTENTS

CHAPTER 1

NEED FOR ENVIRONMENTAL EDUCATION AND PUBLIC AWARENESS TOWARDS SUSTAINABLE DEVELOPMENT

Dr. Dhiraj Kumar Yadav

Senior Assistant Professor, Department of Farm Forestry, UTD
Sant Gahira Guru Vishwavidyalaya, Sarguja, Ambikapur-497001 (C.G.), India
+91-9926615061 E-mail: dheeraj_forestry@yahoo.com

Abstract

Our environment is constantly changing, and we cannot deny that fact. Environmental compassion in our country can only grow through a major public awareness movement. The incredible signification of the message of environmental education for the survival of man has made the relatively new academic discipline to attract great attention. Lack of adequate environmental knowledge is an obstacle in achieving a sustainable future for humankind at both global and local levels. Environmental awareness programme can help reduce the impacts of human activities on the environment. Environmental education has an effective role in creating healthy awareness and preparing suitable environment for the development and maintenance of human minds. It includes both formal and informal education and training that increase human capacity and capability to participate in environmental management and in solving environmental crisis and challenges. This could be achieved by increasing awareness and effectively changing the individual outlook on the environment. Educating adult, community, traditional and religious leaders on the importance of environmental health and on the well use of the natural environment is also very critical to sustainable development. This article is an attempt to review the meaning of environmental education, its aims and objectives, and ways of promoting environmental awareness in the society. It also considered how environmental education can be implemented in many contexts. The article concluded with recommendations of ways of sensitizing the public on the need to be environmental conscious. With a massive influx of natural disasters, warming and cooling periods, different weather patterns, and much more, people need to be aware of the environmental problems our planet faces today.

Keywords: Pollution, Education, Environment, Awareness, Global warming.

Introduction

"A clean and healthy environment is required to achieve the Sustainable Development Goals. Efforts to counter the triple planetary crisis of climate change, nature and biodiversity loss, and pollution and waste must be ramped up in order to truly transform societies and economies for everyone, everywhere."

The Sustainable Development Goals (SDGs) were adopted by all United Nations Member States in 2015 to end poverty, reduce inequality and build more peaceful, prosperous societies by 2030.

The widespread destruction of the natural world is undermining the pursuit of the SDGs and imperiling humanity's future. A human-induced triple planetary crisis of climate change, nature and biodiversity loss, and pollution and waste is pushing nature to the breaking point. That is putting at risk the food we eat, the air we breathe, the water we drink and the materials and resources upon which our societies are built. The triple crisis is weighing most heavily on the vulnerable, including the poor, women and indigenous peoples. Unless humanity reverses its environmental course, it will be impossible to achieve the SDGs.

Attitude towards the environment can be defined as all positive or negative attitudes and thoughts that people show towards behaviours that are beneficial to the environment, such as fears, anger, restlessness, value judgements and readiness to solve environmental problems (Erten, 2005). In order to increase behaviours for the benefit of the environment, positive attitudes should be emphasised. In this context, the importance of interactions with the natural environment is emphasised in order to develop positive behaviours towards the environment (Christine and Mannang, 1997). Pollution, ozone depletion, acid rain, global warming, depletion of natural resources, overpopulation, waste disposal, deforestation, desertification and loss of biodiversity are among the various processes that can be said to cause environmental problems on a global scale (World Health Organization, 1998). Looking at the root cause of environmental problems, it is seen that human behaviour is the main cause of these problems (Tuna, 2000; Vlek & Steg, 2007; Gül, 2013; Singh & Singh, 2017).

"Environmental education is a learning process that increases people's knowledge and awareness about the environment and associated challenges, develops the necessary skills and expertise to address the challenges, and fosters attitudes, motivations, and commitments to make informed decisions and take responsible decisions. Awareness is the basic tool for protection of the environment. Environmental studies in India focus on understanding the complex relationships between human and natural systems, and addressing the various environmental challenges facing the country, such as climate change, deforestation, air and water pollution, and biodiversity loss.

Environmental Studies

Environmental studies can be defined as the study of the interactions between humans and the natural environment. Environmental studies aims to understand, analyze, and solve environmental problems using scientific methods and ethical principles. Environmental studies also foster the environmental awareness, literacy, and citizenship among the people, which are essential for the environmental action and advocacy.

Scope and Importance of Environmental Studies

Environmental studies emerged as a distinct field of study in the mid-20th century, as a result of the growing awareness and concern about the environmental impacts and challenges of human activities, such as industrialization, urbanization, population growth, and consumption. The scope of environmental studies in India is vast and diverse, encompassing a wide range of fields and disciplines. Environmental studies in India focus on understanding the complex relationships between human and natural systems, and addressing the various environmental challenges facing the country, such as climate change, deforestation, air and water pollution, and biodiversity loss. The field includes research and analysis of environmental policies and regulations, environmental impact assessments, sustainable development practices, and conservation efforts. Furthermore, environmental studies in India aim to raise public awareness and education about environmental issues, and to develop innovative solutions to ensure a sustainable future for the country.

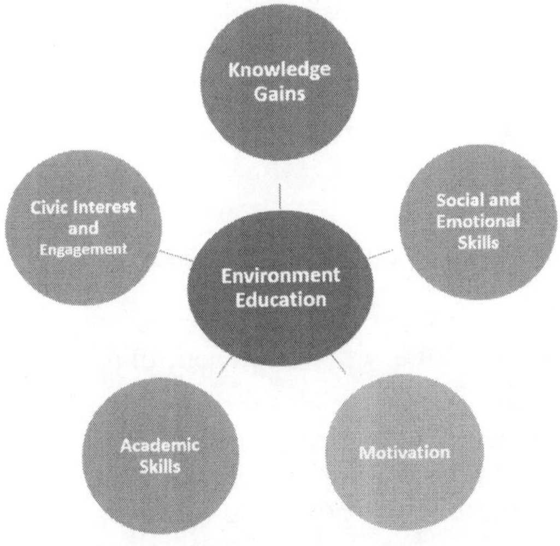

Fig 1: Need and Importance of Environmental Education

Importance of Environmental Studies

Environmental studies is an important and relevant field of study in the present and future setup, as it deals with the environment and its issues, which are crucial for the well-being and survival of humans and other living beings. Environmental studies offers various benefits and advantages, such as:

- Enhancing the environmental knowledge and awareness of the environmental systems and processes, and the environmental impacts and risks of human activities and systems.
- Developing the environmental skills and competencies for measuring, monitoring, and modeling the environmental components and indicators, and for assessing and mitigating the environmental impacts and risks of human activities and systems.
- Fostering the environmental values and attitudes for appreciating and respecting the environment and its diversity and integrity, and for caring and acting for the environment and its protection and conservation.
- Empowering the environmental participation and engagement in the environmental decision-making and governance processes, and in the environmental advocacy and activism movements and campaigns.
- Enabling the environmental innovation and creativity for discovering and developing new and sustainable technologies and practices that can enhance the efficiency and productivity of environmental resources and services, and reduce the environmental footprint of human activities and systems.

These importance of environmental studies in points show that environmental studies is a valuable and useful field of study that can help us to improve our relationship and interaction with the environment and its issues. The Public Environmental Awareness programme aims at creating awareness amongst general public and different segments of the society for minimizing and controlling pollution. Campaign activities are planned and executed under this plan scheme. These include:

- Anti-Littering and Anti Plastic Bag Campaign including waste minimization and management.
- Clean Yamuna Campaign.
- Anti-Fire Crackers Campaign on the occasion of Diwali festival.
- Prohibiting burning of leaves campaign.
- All other environmental & pollution control related issue.

The Department creates awareness by adopting multidimentional approach involving various media. The campaigns also involve active participation of Residents Welfare Associations, Market Traders Associations, Industrial Associations, Schools/Colleges, Institutions, Corporates & NGOs.

The role of education in environmental awareness

Education can help people acquire the knowledge, skills, values, and attitudes that enable them to understand and appreciate the natural environment and its interconnections with human activities. Education can also foster critical thinking, problem-solving, and creativity, which are essential for finding innovative and sustainable solutions to environmental challenges. Moreover, education can promote civic engagement, ethical responsibility, and global citizenship, which can motivate people to take action and participate in environmental movements and initiatives.

The types of education that matter for environmental awareness

Not all forms of education have the same impact on environmental awareness and behavior. Some studies suggest that formal education, especially at higher levels, can increase environmental literacy and concern, but may not necessarily translate into pro-environmental actions. Other studies indicate that informal and non-formal education, such as experiential learning, community involvement, media exposure, and social networks, can have a stronger influence on environmental behavior, especially when they are relevant, meaningful, and empowering for the learners. Therefore, a combination of different types of education, tailored to the context and needs of the learners, may be more effective for enhancing environmental awareness and behavior.

The hurdles and opportunities for environmental education

Despite the potential benefits of education for environmental awareness and behavior, there are also many barriers and challenges that limit its effectiveness and reach. Some of these include: lack of resources, support, and incentives for educators and learners; inadequate curricula, pedagogy, and assessment; resistance and skepticism from some stakeholders and groups; and competing priorities and interests in the economic and political spheres. However, there are also many opportunities and initiatives that can overcome these barriers and promote environmental education, such as: collaboration and partnership among different sectors and actors; integration and innovation in the educational system and practice; advocacy and awareness-raising among the public and policymakers; and recognition and reward for environmental achievements and contributions.

The impact of education on environmental outcomes

The ultimate goal of education for environmental awareness and behavior is to contribute to positive environmental outcomes, such as reducing greenhouse gas emissions, conserving natural resources, protecting biodiversity, and improving human well-being. However, measuring and attributing the impact of education on these outcomes is not easy, as there are many other factors and variables that influence them. Moreover, the impact of education may not be immediate, direct, or linear, but rather long-term, indirect, or complex. Therefore, a holistic, multidimensional, and adaptive approach is needed to evaluate and improve the impact of education on environmental outcomes.

The future of education for environmental awareness and behavior

The current and future environmental challenges require a transformation of the way we think and act towards the environment. Education can play a vital role in facilitating this transformation, by providing the foundation and the tools for environmental awareness and behavior. However, education alone is not enough, as it also depends on the social, cultural, economic, and political context and conditions that shape and constrain it. Therefore, education for environmental awareness and behavior needs to be part of a broader and more integrated strategy that involves multiple stakeholders and sectors, and that aligns with the principles and goals of sustainable development.

Environmental Awareness

To explain environmental awareness, we must first understand the environmental movement. Environmentalism is an ideology that evokes the necessity and responsibility of humans to respect, protect, and preserve the natural world from its anthropogenic (caused by humans) afflictions. Environmental awareness is an integral part of the movement's success. By spreading awareness to others that the physical environment is fragile and indispensable, we can begin fixing the issues that threaten it.

How to Endorse Environmental Awareness

Before we can begin promoting environmental awareness in your community, we must first make sure that you have a thorough understanding of environmental issues. Stay up to date on environmental news, read books and other materials, and learn about the issues affecting your own community. It's much easier to talk to others about the environment if you've already taken the time to educate yourself.

Numerous resources are available to promote environmental awareness and education: group learning (inside or outside of the classroom), informational and inspirational seminars, online courses, books, articles, videos, and brochures. These are just a few of the tools that can get you started in raising awareness about the environment.

Examples of Environmental Issues

- **Pollution:** Pollution of the air, water and soil caused by toxins such as plastics, heavy metals and nitrates, caused by factors such as toxins and gases released by factories, combustion of fossil fuels, acid rain, oil spill and industrial waste.
- **Global warming:** The emission of greenhouse gases due to human activity causes global warming, which in turn causes an increase in temperature that then leads to rising sea levels, melting of polar ice caps, flash floods and desertification.
- **Overpopulation:** We are facing a shortage of resources such as food, water

and fuel to sustain the rising global population, particularly in developing countries. Intensive agriculture attempting to lessen the problem actually leads to more damage through the use of chemical fertilizers, pesticides and insecticides.

- **Waste disposal:** An excessive amount of waste is produced and dumped in the oceans. Nuclear waste is particularly dangerous, as well as plastics and electronic waste.
- **Ocean acidification:** The increase in the production of carbon dioxide by humans causes the oceans' acidity to rise, which has a negative impact on marine life.
- **Loss of biodiversity:** Species and habitats are becoming extinct due to human activity. This causes an imbalance in natural processes like pollination and poses a threat to ecosystems – coral reef destruction is particularly affected.
- **Deforestation:** Loss of trees in order to make space for residential, industrial or commercial projects means that less oxygen is produced, and temperature and rainfall are affected.
- **Ozone layer depletion:** Pollution caused by chlorofluorocarbons (CFCs) in the air creates a hole in the ozone layer, which protects the earth from harmful UV radiation.
- **Acid rain:** Pollutants in the atmosphere such as sulfur dioxide and nitrogen oxides cause acid rain, which has negative consequences for humans, wildlife and aquatic species.
- **Public health issues:** Lack of clean water is one of the leading environmental problems currently. Pollutants in the air also cause issues such as respiratory disease and cardiovascular disease.

Let's Make a Transformation

Once we are well versed in environmental concerns, we can use that information to start beneficial projects in our home and/or in our community.

Here are some examples of the kinds of projects and initiatives you could get involved in:

- Instead of driving to work or school, take the bus, carpool, walk, or ride your bike to cut down on greenhouse gas emissions. According to the EPA transportation adds to 33% of the total atmospheric CO .
- Consider investing in appropriate technology like clean power (solar or wind), if not for your home then maybe for a community center. This supports a transition to clean and renewable energy.
- Buy reusable products such as glass bottles, reusable bags, and reusable cups. Avoid buying disposable goods such as paper towels, plastic bottles, and plastic bags.

- Start a composting and recycling program in communities where there are none. This will help cut down our waste production.
- Support local businesses and farmers, and buy organic and pesticide-free food when you can. Or, start your own community garden.

Once you have chosen your cause and have started a personal or community project, share it with the world! Get your coworkers, neighbors, friends, family, or even your local government involved. It's much easier and more effective to spread environmental awareness and start a local project if you collaborate with others in your community. Promoting environmental awareness is a crucial part of being an environmental steward. Start participating in the change and teach your community what is needed to create a sustainable future.

"Deepen your environmental awareness by learning about environmental rights"

Government of India has passed several Acts for the protection of environment. The most important ones are –

- Wild life protection Acts, 1972-1986.
- Water (prevention and control of pollution)Acts, 1974.
- Water pollution less Act, 1980.
- Forest (conservation) Act, 1980
- Air (prevention and control of pollution) Act, 1981
- The Environment (protection) Act, 1986.
- Biological Diversity Act, 2002

Fig 2: Environmental Sustainability

Creating an environment to educate about the environment

The environment is becoming increasingly contaminated by our imagination. Most human activities cause the environment to become contaminated. This will affect both the global and regional levels. Thinning of the ozone layer and the increase in greenhouse gas emissions are examples of global losses. While water pollution, soil erosion is one of the few regional outcomes created by human activities and by them the environment is also affected. Environment education and awareness is comprehensive, continuing process, rooted in broad interdisciplinary base. Environment education is multifaceted and touches upon every aspect of life. It aims to assist individuals an understanding that people are inseparable parts of an environmental system and that whatever they do alter their surroundings in both harmful and beneficial ways. Therefore, our voice is essential for environmental education for environmental protection and management. It is a right step to teach people and society the best way to utilize present and future resources. Through environmental education, all people can gain knowledge of correcting the fundamental issues leading to local pollution.

What is Environmental Education (EE)?

Environmental education is a multi-disciplinary approach to the study of man's problems of maintaining a livable Earth. Technically, environmental education is an expansion of study of ecology, a branch of biology dealing with inter-relationship of living organisms with one another and their surroundings. In environmental education, people are told that how to manage the ecosystem in order to maintain the environment and the pollution free environment. Environmental education provides the necessary skills and specialized knowledge to meet the challenges related to it. The main purpose of this education is to provide knowledge, to create awareness, to create an attitude of contemplation and to provide the necessary skills to control the environmental challenges. Environmental Education after the Stockholm Conference on Human Environment organized by UNESCO in 1972 had gained reputation on a global scale. The concern for the environment in the present era found expression in the proclamation made to 'Defend and improve the environment for the present and future generations' by United Nations. Immediately after this conference, UNESCO also launched the International Environmental Education Program (IEEP). Since then, the environmental problems have attracted the attention of a world cross- section of people all over the world. Objectives of Environmental Education The inter-governmental conference on environment education held at Tbilisi, USSR (1977) listed the following objectives of environmental education.

- **Knowledge:** To help people acquire basic understanding of the total environment and associated problems.
- **Attitudes:** To help people acquire social values, strong feelings of concern for the environment and the motivation to act in protecting and improving it.

- **Awareness:** To help people acquire an awareness of and sensitivity to the total environment and its allied problems.
- **Skills:** To help people to acquire skills for solving such problems and to evaluate environmental measures.
- **Participation:** To help people develop a sense of responsibility and urgency regarding environmental problems and solutions.

Necessity of environmental education

Each country is engaged in efforts to solve environmental concerns with education. According to these different countries, environmental education should not only be a part of the education system, but also participate in the political system, so that the work, policies and appropriate plans can be prepared at the national level. Environmental education should be able to assess the environmental condition and play a leading role in reducing environmental damage. It should be planned according to the rules, because normal changes in daily life can make a huge contribution in improving the environment. Environmental Education and its Relevancy Environmental education shall have relevance to the following aspects;

- Public awareness
- Identifications of environmental problems and solutions
- Conservation and preservation of natural resources
- Sustainable development
- Successful and effective planning of national economy

Security of the environment is the responsibility of everyone. Therefore, environmental education should not be confined to a group or society, but every person should be aware of environmental protection. It should be a continuous and lifelong process and should be pragmatic towards environmental education so that it can be implemented properly. Environmental Awareness Environment awareness is the ABC of environmental literacy, environmental education, environmental issues, and environmental programs. It is the first step of the ladder to the environmental security for sustainable development. We all know the environment is close to linked to human beings and hence influences their health and quality of life. So improving our environment by environmental awareness and education will improve the human health and well-being of all.

In Indian universities, there has been considerable emphasis on education, research and training. Different universities and institutions are imparting courses like Environmental Engineering, Conservation and Management, Environmental Health and Social Science. Importance of environmental awareness Environmental Awareness leads to understanding the environmental issues. Helps implementation of environmental programs, laws etc. creates concern about environmental impact on Flora and Fauna and well-being of man as well. If the children are taught about the resources, environmental pollution, soil erosion, deflation and the protection and protection of endangered plants and extinct animals, the protection of the environment can be substantially improved. Education is a type of investment that changes over time as a valuable asset.

In order to promote nationwide environmental awareness, Government of India had established the Environmental Education Center (CEE) in August 1984 with the support of Ministry of Environment and Forests. One of the main functions of CEE is that efforts should be made to give proper recognition to the role of environmental education. CEE runs many educational programs related to it.

Environmental education strategies

The task of achieving environmental education goals calls for some special attention to selecting effective educational tactics. Education – teaching and learning- is the logical starting point for any cultural effort targeting the public. Education is the basic factor for achieving sustainable development and improving the individuals' skills to cope with development and environmental issues. According to Peyton et al. (1997) some recommendations can be offered to guide teachers and curriculum developers in designing approaches to environmental education.

- **Incorporating environmental issues in school curriculum:** It is a process of integrating environmental education concepts, skills and strategies throughout an existing general education curriculum, instead of a mono disciplinary model in which environmental education is treated in a separate course or as a comprehensive unit within a course. The infusion of environmental education would enhance existing programmes without competing for limited curriculum time and resources. Social sciences, psychology, economy, geography, history together with information technology could be used to develop and implement solutions for complicated problems. Environmental education should also be pervaded within teacher training- programme.
- **Scheming environmental awareness programmes:** With cooperation between different mass media (TV, radio, newspaper, posters etc) a well-designed awareness programmes can be conducted to the society to disseminate knowledge about the environment and its problems and the role of community participation in protecting the environment.
- **Exhibitions and Displays**: Dioramas representing the different ecosystems in the country including the fauna and flora especially threatened species of animals and plants are effective educational tools. The Natural History Museum in the University of Khartoum could play a major role in this task due to its mission to disseminate the environmental message to a broad spectrum of the community from a pre-school child to a mature scientist, in addition to its contribution in national and special celebration activities. Indeed, it has been pursuing this for the last three decades. The IES also has a newsletter "Sudan Environment" and "*Environmental Monography*" series.

- **Establishment of environment clubs and societies:** Activities on environmental issues conducted in social and cultural clubs by individuals and social groups such Ecofriends, Consumer Protection Society and Sudanese Environment Conservation Society are playing a leading role in environmental education. Lectures, seminars, posters, exhibitions or any activity of concerning environmental problems can suitable tools for such activities.

World Environment Day being celebrated on 5th of June across the world. The Kingdom of Saudi Arabia hosted the 2024 World Environment Day global celebrations. The theme 2024 states, *"Land Restoration, Desertification, and Drought Resilience."* It's tied in with bringing back healthy land, keeping desserts from developing, and managing water shortages. Trees, healthy soil, and clean water are imperative for a solid planet. The drive focused under the slogan *"Our land. Our future. We are #GenerationRestoration."*

Conclusion

The need for protection and conservation of environment and sustainable use of natural resources is reflected in the constitutional framework of India. The governance for environmental management broadly comprises of legislative framework, policies and a set of regulatory institutions. India has a vast network of such laws, policies and institutions too. If we continue moving forward in such a harmful way towards the future, then there will be no future to consider. Raising awareness in our local community and within our families about these issues can help contribute to a more environmentally conscious and friendly place for you and your future generations to live. Environmental studies are a multidisciplinary and holistic field of study that systematically studies the environment and its issues from a human-oriented perspective. In this article we discussed the meaning, importance and scope of environmental studies, reasons to study environmental studies. Environmental studies integrates knowledge from the natural sciences, the humanities, and the social sciences to understand, analyze, evaluate, and address the causes, consequences, and solutions of environmental problems. Environmental studies also explore the values, ethics, and policies that guide the human-environment relationship and the environmental decision-making and governance processes.

Environmental education and environmental awareness programme are to be conducted from primary to university level of education are the better channels to which information can be disseminated to the grassroots level and these programmes would enable the masses to protect the environment which is very urgent for us to survive on planet. Environmentally illiterate individuals cannot successfully solve complex problems. There is an urgent need to safeguard our environment to save the future generation. Environmental awareness is the suitable tool for conservation of our natural resources and their valuable components. Thanks to many environmental *Think-Tanks* and organizations like UN, awareness is on the rise. And this has encouraged many people to make positive changes for the conservation and protection of the environment

References

Acharya, K.P., Paudel, P.K., Neupane, P.R., and Köhl, M. (2016). Human-Wildlife Conflicts in Nepal: Patterns of Human Fatalities and Injuries Caused by Large Mammals. *PLoS One* 11:e0161717. doi: 10.1371/journal.pone.0161717

Ahluwalia, S.P. and Bais, H.S. (1992). Environmental education. Ashish publishing house, New Delhi.

Anonymous. 1983. Trends, needs and priorities of environmental education since the Tbilisi Conference an overview. Environmental Education Series No. 1. UNESCO-UNEP International Environmental Programme. Paris, France. 43 pp.

Bhaskaracharyulu, Y. (2011). Education and Society. Discovery Publishing house, New Delhi. 20 pp

Buragohain, P.P et al. (2017). Paribesh Adhyayan, Banalata, Dibrugarh.

Christine, N., & Mannıng R.(1997). Incorporating environmental behavior, ethics and values into nonformal environmental education programs. *Journal of Enviromental Education,* 28(2), 10-21. http://dx.doi.org/10.1080/00958964.1997.9942818

Dutta,, B.(2023). Environmental Education and Public Awareness. *Journal of Humanities and Social Sciences*. Vol. III (1) 2023.

Erten, S. (2005). Investigation of environmentally friendly behaviours in pre-service preschool teachers. *Hacettepe University Journal of Faculty of Education,* (28), 91-100.

Goswami, M. (2016). Environmental education and Awareness.

Gül, F. (2013). Environmental problems and philosophy in the context of human-nature relationship. *Pamukkale University Journal of Institute of Social Sciences,* 14, 17-21.

Hungerford, H. R. and T. L. Volk. 1990. Changing learner behavior through environmental education. The Round Table on Environment and Education at the World Conference on Education for All. Sponsored by UNESCO, UNICEF and UNEP. Jomtien, Thailand. March. 28 pp.

Nath, Bhaskar (2009). Environmental education and Awareness. Vol.1. Encyclopedia of Life Support System.

Peyton, B., Campa. H., Winterstein, S. R., Peyton, M. and Peyton, J. 1997. Biological Diversity for Secondary Education. Environmental Education Module. UNESCO- UNEP. International Environmental Education Programme (IEEP). 153 pp.

Rokeach Milton. 1968. Beliefs, attitudes and values: a theory of organization and change. Josssey- Bass, Inc., Publishers. San Francisco, Calif. 214 pp.

Singh, R. L. & Singh, P. K. (2017). Global environmental problems. *Principles and Applications of Environmental Biotechnology for A Sustainable Future*, 13–41.

Tuna, M. (2000). Globalisation of environmental problems. *Muğla University Journal of Institute of Social Sciences* (2), 1-16.

United Nations Educational, Scientific and Cultural Organization and United Nations Environment Programme., 1978. The Tbilisi Declaration. Connect. UNESCO- UNEP Environmental Education Newsletter. January, 111(1) 8 pp.

United Nations Educational, Scientific and Cultural Organization and United Nations Environment Programme. 1987. International strategy for action in the field of environment United Nations education and training in the 1990's. UNESCO, 7, Place de Fonteyn, 75700, UNEP, Nairobi, Kenya. 21 pp.

Vaishali Anand (2020). Environment and Ecology. McGraw Hill Education (India) Private Limited. 468 pp.

Vlek, C., & Steg, L. (2007). Human behavior and environmental sustainability: Problems, driving forces and research topics. *Journal of Social Issues*, 63(1), 1-19.

CHAPTER 2

CONTRIBUTION OF ELECTRIC VEHICLES IN THE TRANSPORTATION SECTOR TO PROTECT THE ENVIRONMENT

Demendra Kumar Ravi[1], Dr. Gyanendra Shukla[2], Dr. Ritu Marwah[3]

[1]*Research scholar, Govt. D.B.P.G. Girls College, Raipur, C.G.*
[2]*Asst Prof, Govt College Birgaon, Raipur, C.G.*
[3]*Asst. Prof., Govt. D.B.P.G. Girls College, Raipur, C.G.*
demendra.ravi@gmail.com

Abstract

The environment is the only thing that is automated, well-managed, and organized several decades ago when there was less interference from humans, no luxurious innovation, fewer human populations, and less wastage. This environment was at its best level by itself, because of balanced climate conditions, clean water, air, and soil quality, balanced eco-system, greenery everywhere, and no harm to the ozone layer but after getting knowledge and power in human hands, we started innovating things like fuel engines operated vehicles, Air-conditioner, refrigerator, microwave own, geyser, iron, heater and started exploiting forest to make homes by concrete and rock. We humans are snatching away the shelter of millions of animals to get our dream homes. All these environmental issues are at a peak level which is why humans need to adopt sustainable technology to manage a smart environment. There are multiple options for sustainable technology but we have taken electric vehicle contribution as a sustainable option to protect our environment. In this article, we have also discussed the role of the transportation sector in our environment.

Keywords: Electric vehicle adoption, Environment, Sustainable technology, Transportation sector.

Introduction

It is time to act and make needful changes by adopting sustainable technologies to protect our environment from climate change and various environmental pollution. We must protect the environment. We know that technology is developing year after year to meet the expected and unexpected needs and wants of humans to make living easy. This technology has both good and bad characteristics. However, it is not only about technology and innovation but also a matter of choices which is directly and indirectly associated with human behaviours.

According to (Rodgriue, 2017) transportation sector played two roles one positive and the other negative, on the positive side it supported mobility for passengers, and on the negative side, this sector impacted the environment negatively. The transportation sector contributes 25% to global CO2 emissions and it is 28% for a developed country like the United States. The transportation sector has a major contribution to environmental pollution. Petroleum engine vehicles like all kinds of 4-wheelers, 3-wheelers, and 2-wheelers are used as the most preferred and easiest mode of road transportation. According to Fortune Business Insights (Business Insights, 2024) The global market size of two-wheelers is valued at USD 116.82 billion in 2023. The projected market growth is at USD 123.54 billion for 2024 and it will grow at USD 215.96 billion by 2032. The projected growth is at a CAGR of 7.2% from the period 2024 to 2032. The global two-wheeler market includes manufacturing, distribution, and sales, despite uncertain fuel prices and economic uncertainties two-wheeler market will continue to grow. The electric two-wheeler market is also growing due to its connected features and environmental awareness. We understand that the environment is being impacted heavily by the transportation sector due to its harmful carbon emissions and here a question arises What should we do? How to protect the environment? And the answer is to adopt sustainable technology like Electric Vehicle in the transportation sector. Healthy environmentally friendly moves by every person will also help to protect the environment.

Effect on the environment

According to the article "Climate Change Impacts" by (NOAA, 2021) climate change has impacted multiple things such as below:

- Global temperature rose about 1.8 F (1 C) from 1901 to 2020.
- Sea level rise has accelerated from 1.7 mm/year throughout most of the twentieth century to 3.2 mm/year since 1993.
- Glaciers are shrinking: the average thickness of 30 well-studied glaciers has decreased more than 60 feet since 1980.
- The area covered by sea ice in the Arctic at the end of summer has shrunk by about 40% since 1979.
- The amount of carbon dioxide in the atmosphere has risen by 25% since 1958, and by about 40% since the Industrial Revolution.
- Snow is melting earlier compared to long-term averages.

The adoption of sustainable technology is a complex interplay of psychological, social, and cultural factors. Understanding these influences can help stakeholders develop more effective strategies to encourage widespread adoption. By addressing psychological biases, leveraging social norms, respecting cultural values, and implementing practical measures, we can facilitate a smoother transition to a more sustainable future. As we continue to innovate and seek solutions to pressing environmental challenges, a nuanced understanding of human behaviour will be crucial in driving meaningful and lasting change.

Features of sustainable technology

Sustainable technology is being invented to create balance in the environment and decrease dependency on non-renewable resources. There are multiple features of sustainable technology as below:

a. **Reduce carbon emission:** The feature of reducing carbon emission is important features to protect the environment because increased carbon is very harmful to the environment. According to (Nunez, 2019) greenhouse gas is a blanket of earth, There are multiple gases included in greenhouse gas, and one of GHS is $CO_{2.}$. Greenhouse gas decreases the heat and makes the earth habitual for the living of humans and animals but CO_2 is at record high levels which is out of balance It results in global warming and climate change. That is why we need to reduce carbon emissions by shifting toward Sustainable technologies that reduce carbon emission.

b. **Improve Renewable energy dependency:** At present dependency on non-renewable energy is higher than renewable energy. According to (Counts, 2024) 83% of the global energy is consumed from fossil fuels which are non-renewable. By 2040 global energy consumption will reach 740 million. From 2000 to 2040 there will be a 77% increase in global energy consumption. Renewable energy consumption is on a slow rise. Between 2011 to 2030 renewable energy is expected to grow by 2 to 6% and this renewable energy growth is not enough. So we need to change our energy source as soon as possible. The alternate renewable energy options are solar energy, wind energy, hydro energy, geothermal energy, and biomass energy. It is one of the features of sustainable technology.

c. **Energy efficient:** There are multiple sources of energy but every energy source that we use for something has limitations. For example, in transportation a petroleum car has a fuel capacity of 45 litters runs 12 KM/L can be driven 540 kilometres, and approx. fuel cost would be 4500 INR but when we consider an electric car having 35 kWh renewable battery can give a range of 400 kilometers on a single charge. The cost of an electric car ride is not more than 500 INR for 400 kilometers. This electric car can be charged again at home. For energy efficiency it includes operational energy consumption should be less, operational cost should be reduced, environment friendly, and performance outcomes should be the same or more.

d. **Reusable and recyclable:** There is the absence of reusability and recyclability features in non-renewable energy sources but when we consider renewable energy sources it includes both reusability and recyclability features. It aimed to reduce wastage and increase the lifecycle of products. In further this feature is also known as a complete product separated into categories like iron, aluminum, and plastic being broken down or melted into small pieces and these pieces became raw materials for new product creation. There are multiple products we use that have these features online application instead of offline application, digital copy instead of text copy, and rechargeable battery instead of petroleum.

e. **Durable and longevity:** The creation of a product should follow the long product life cycle without compromising the performance of that product. It should be tested in multiple situations. After the consumer uses it should not need frequent repair and replacement which will help to save environmental resources and create less wastage. In short durable means less degradation and longevity means relevant and effective over time.

f. **Reduced environmental impact:** It means minimizing the negative impact on the environment by waste control of the product, reducing noise and air pollution, reducing carbon emission, the decline in non-renewable energy dependency, and less resource use, all this practice together reduces the negative environmental impact

g. **Smart and adaptive technology:** A good example is an electric vehicle, it includes energy efficiency, and smart connectivity features such as phone connectivity, app support, navigation system, auto-hold, reverse mode, and many more. These features are very attractive among consumers and help in sustainable technology adoption.

Electric vehicles and human behaviour

The contribution of the transportation sector is very high to affect the environment because no one can stay at home all the time, Humans must go out to bring food, do the job, meet friends and relatives, and like to go out to explore new places, humans are social animals, and they have multiple needs, wants, and desires. and for fulfilling all these things required transportation facility. The innovation in the transportation sector has continued to make things possible. Previously, there were only diesel, petrol, and CNG fuel options available for combustive engine vehicles. Diesel-engine vehicles create more air pollution than petrol-engine vehicles which is why the use of Urea in diesel-engine vehicles is mandatory at present. Urea is a chemical compound that reduces nitrogen oxide to reduce acid rain and air pollution but that is not enough. There was only combustive engine vehicles were available to cover the distance by roads only when the transportation sector was untouched by sustainable technology, a new option of electric vehicles came into existence to protect and transform this sector to the next level. Electric vehicles are getting attention from consumers and the government. It is useless if sustainable technology is developed but people are not shifting towards sustainable technology on time. It is not beneficial if people are not adopting electric vehicles at the right time. Human buying behaviour of electric vehicles is affected by two major aspects: psychological such as attitude, perception, and behaviour. It is psychological where a person behaves for something based on his knowledge, society, and culture and acts accordingly and the other effect is secondary and very materialistic such as features, cost, and performance. Concern about climate change and carbon emission should be primary while being vehicles and habitual of riding petroleum vehicles should be secondary. People buy things by getting information and recommendations from their known people and good marketing. Adoption of a product is a process where customers start accepting a product according to their needs and usefulness, it is not a one-time event but a continuous process that can take several months and years, and then customers become familiar with using the product. Electric vehicles in India are also at their early stages, due to lake of charging infrastructure but this can be changed by improving charging infrastructure. As we can see in Chart 1 there are 4 levels of adoption:

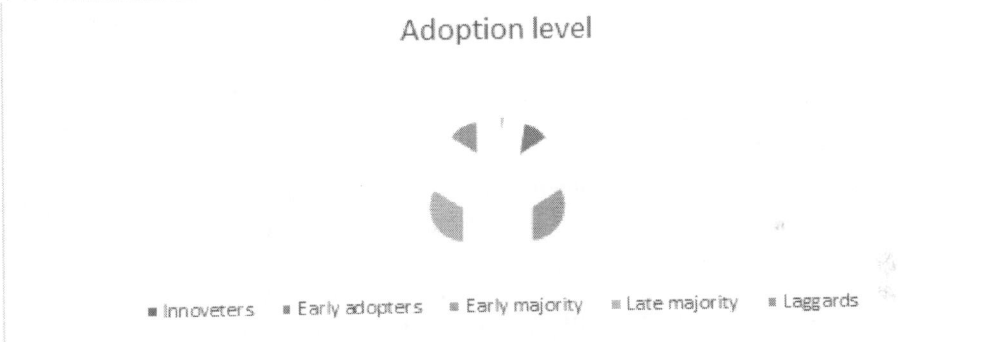

Fig 1: Level of Adoption

Advantages of having electric vehicles

a. Riding costs are pocket-friendly, so customers can save their money by adopting electric vehicles.

b. Increased carbon emission is the main concern all around the world and electric vehicles can help to reduce carbon emissions, which will create a healthy environment by reducing acid rain and global warming.

c. Multiple feel-good features come with electric two-wheeler vehicles like reverse mode, phone connectivity, inbuilt sim, tracking of carbon-free ride and some safety features also come with electric vehicles like auto-hold, and fall detection. There is no comparison between electric 4-wheelers and petroleum engine 4-wheelers because all advanced features come in both types of 4-wheelers.

d. Electric vehicles do not need regular costly servicing because less component is used in electric vehicles but in petroleum engine vehicles various components like oil filters, oil, air filters, and large-size engines are used.

e. The government is in favor of increasing the dependency on electric vehicles because non-renewable energy is available in limited quantity, expensive, and harmful to the environment. Indian government imports this crude oil from other nations and this is one of the reasons for the deficit balance of payment, so the concept of electric vehicles is a good substitute for petroleum engine vehicles to balance non-renewable energy sources. The government provides subsidies and decreases road transportation office (RTO) costs to promote electric vehicle buying and running various schemes like only electric vehicles on the road till 2030, special electric mobility zone, battery switching scheme, tax deduction, and many more.

f. Electric vehicles are not only pocket-friendly but also environmentally friendly due to their various features like charge and ride, fewer components used, and help to reduce carbon emission.

Barriers to shifting toward electric vehicles

There are multiple challenges associated with the electric vehicles that need to be discussed. These barriers play an important role in the decision-making of buying electric vehicles among customers.

a. The battery capacity of electric vehicles is very important when buying a new electric vehicle, and as per battery capacity, the vehicle covers a distance in km on average, if the battery size is small range coverage will be less and for increased battery capacity it will give good range and very costly. There are multiple modes on it like eco and sport, But the problem is that very few people run on eco mode because this mode compromises the riding performance in terms of speed and pick-up. So, most people want good riding performance on highways.

b. After long-term use of electric vehicles battery performance starts declining and needs to be changed and it is very costly.

c. There is a lack of good charging infrastructure in India, therefore people cannot think about going anywhere on electric vehicles. Fuel stations are available everywhere but charging stations are available in very few numbers and conditions are very poor in rural areas.

d. The cost of electric vehicles is higher than petroleum engine vehicles and there are fewer options for electric vehicles available in the market. Due to this petroleum engine vehicle sales are more than electric vehicle sales at present.

e. There are multiple news that have taken place among customers that electric vehicles catch fire while charging or driving electric vehicles due to short circuits or overheating of the charger or vehicle.

f. There is a major issue of lack of service or repair centers available. If any problem arises in electric vehicles, it can be only repaired in authorized service centers.

Suggestions

- An increase in public charging availability in most places, like fuel stations, will build consumer's confidence in the rapid shift to electric scooters. Example: Jio Telecom started.
- The charging time at home should be reduced.
- Subsidy should be available as soon as possible; the government should investigate this and act against this time-taking subsidy process.
- The ownership cost of an electric scooter is a little high; companies should reduce the cost.
- Battery replacement costs are 50% of the cost of vehicles. Proper planning is needed.
- Electric scooter companies should increase the number of service centers and employees. And run a tie-up program with local service centers.
- Safety should be a top priority, and the government should make strict rules for the electric scooter industry.
- Solar energy should be used in public charging stations to charge electric vehicles.
- The government should encourage public transportation facilities in large size in Chhattisgarh state.

Conclusion

Indian customers mostly prefer cars, buses, trucks, and bikes on the road for going anywhere which is why road transportation has a major influence on the environment. Most of the users have not yet shifted to electric vehicles, but there will be a good shift toward the electric vehicles segment in the near future. We have not several centuries in our hands to save nature. Humans need to understand that it is only their responsibility to save the environment, not any magic will happen automatically. Initiative of the electric vehicle shift can help a lot to manage our environment, there are multiple types of electric vehicles electric cars, electric trucks, electric buses, electric rickshaws, and electric scooters. Electric scooters are more popular than other electric vehicles. The government should control the cost of electric vehicles so that people can get this easily.

References

Business Insights, F. (2024, July 15). Two-wheeler Market Size & Share | Growth Analysis [2032]. https://www.fortunebusinessinsights.com/two-wheeler-market-106884

Counts, T. W. (2024). Global Energy Consumption. https://www.theworldcounts.com/challenges/climate-change/energy/global-energy-consumption

NOAA, E. (2021). Climate change impacts | National Oceanic and Atmospheric Administration. https://www.noaa.gov/education/resource-collections/climate/climate-change-impacts

Nunez, C. (2019, May 13). Greenhouse gases, facts, and information. Environment. https://www.nationalgeographic.com/environment/article/greenhouse-gases

Rodgriue, Dr. J.-P. (2017, December 7). Transportation and the Environment | The Geography of Transport Systems. https://transportgeography.org/contents/chapter4/transportation-and-environment/

CHAPTER 3

STUDY OF CERTAIN FOREST TREE SEEDS AS A SOURCE OF HUMAN NUTRITION

Dr. Saileshkumar Prajapati

Department of Botany, Government Science College,
Santrampur, Dist. Mahisagar, 389260 (Gujarat)
E. mail: psailesh4@gmail.com

Abstract

Food the most important, fundamental need, evolved with the existence of the living world. Besides the edible seeds which are consumed by the living world, forest tree seeds may be a good source of food. These seeds are available in abundance in the forest but are underutilized because of the ignorance of their nutritional status. Thus,22 forest tree seeds from the Victoria Park (reserved forest) near Bhavnagar(Gujarat) were selected for the study and analysed for their protein, reducing and non-reducing sugars, starch and oil content. From the estimation for biochemical constituents of seeds, out of 22 species 17 species (9 leguminous and 8 non – leguminous) had fairly good amount of protein. The seeds of 16 (10 leguminous and 6 non-leguminous) species found rich with the good amount of total sugar. Seeds of 14 species (9 leguminous and 5 non-leguminous) species found enriched with starch content. In case of oil content seeds of 10 non-leguminous species and seeds of 2 leguminous species had fairly good percentage of oil.

Keywords: Seeds, Nutrition, Forest trees.

Introduction

From the time immemorial food, shelter and clothing are the three fundamental and basic need of human society; among these food is the most important need for the existence of the human beings. Committed over 10,000 years ago, when he learn to satisfy his hunger by growing food, specially seed food. These seeds with their nutritive reserve food and seedling offer the living world a great staple food. Besides edible seeds which are consumed by the living world, forest tree seeds may be a good source of food. These seeds are available in abundance in the forest are under utilized because of ignorance of their nutritional status. Thus, proper awareness required toward the forest tree seeds. In view of the growing incidence of malnutrition, it is imperative that more efforts made to identify locally available tree species, different edible parts of which can be consumed by the forest dwellers to improve the nutrient status of their diet. At such juncture of great anxiety for research of new food commodities the attention naturally diverted towards the forest tree seeds which remained unexploited so far. With these points of view seeds of some tree species of Victoria Park reserved forest were analysed for their protein, reducing, non-reducing and total sugars, starch and oil content.

The studies on seed bio-chemicals were reported in tree species by Bonner (1973), according to him, in most seeds, complex carbohydrates, fats, oil and protein usually accumulates with maturation of seeds. Previously Gupta et al (1995); Soni (1995); Tomar et al (1996); Kadam et al. (1996); Shrivastava and Jha (1997); Agboola and Kadiri (1998); Kadam (2000a & b); Kadam (2001); Bukhari (2002) have studied the biochemical make-up of forest tree seeds.

Materials and Methods

Mature seeds were collected from the reserved forest (Victoria Park) for the biochemical analysis. The reserved forest (Victoria Park) is a dry deciduous forest, situated at 3 km south of Bhavnagar city, which is located at $21^0 42'$ to $21^0 48'$ North and longitude $72^0 04'$ to $72^0 14'$ East on the western coast of the Gulf of Cambay in Peninsula of Saurashtra in Gujarat state. It is semi-arid region with maximum temperature 43^0 C in May and minimum 10^0 C in January. The annual average rainfall is 700 mm, which is irregular and scanty. The entire area is triangular in shape and covers about 202.74 hectares. Seeds were analysed for their protein, reducing and non-reducing sugars and starch content. Following species (11 leguminous and 11 non-leguminous) were selected for the study: *Acacia nilotica* (L.) , *Albizia lebbeck* (L.), *Bauhinia purpurea* (L.), *Butea monosperma* (Lam.), *Cassia fistula* (L.), *Delonix regia* (Boj.), *Derris indica* (Lam.), *Gliricidia sepium* (Jacq.), *Leucaena leucocephala* (Lam.), *Peltophorum pterocarpum* (DC), *Pithecellobium dulce* (Roxb.), *Adansonia digitata* (L.), *Aegle marmalose* (L.), *Ailanthus excelsa* (Roxb.), *Azadirachta indica* (A. Juss.), *Gmelina arborea* (Roxb.), *Mimusops elengi*(L), *Moringa oleifera* (Lam.), *Melia azedarach* (L.), *Sapindus laurifolius* (Vahl.), *Thespesia populnea* (L.), *Sterculia foetida* (L.)

The dried seeds were powered and analysed as per the following methods:

1. **Sugar**: Reducing and non-reducing sugars from the seeds, analysed as per the method given by Nelson-Somogyi (1944). The difference between total sugar and reducing sugar without hydrolysis corresponds to the quantity of non-reducing sugars. (Mahadevan and Sridhar, 1974)

2. **Protein**: Protein estimation was made as per the method given by Folin-Lowry (1951).
3. **Starch**: Starch was calculated by converting the total sugar content by multiplying with conversion factor 0.90 (Mahadevan and Sridhar, 1974)
4. **Oil**: Soxhelt – extractor (Singh, 1987) method was followed for the oil extraction.

Results and Discussion

On the examining the leguminous seeds for their protein content it was found that *Gliricidia sepium* had the fairly good amount of protein while lowest protein content reported form the *Acacia nilotica*. The highest amount of reducing sugar recorded from *Albizia lebbeck* and lowest from *Leucaena leucocephala*. The amount of non-reducing found highest form the *Butea monosperma* while lowest reported from *Gliricidia sepium*. The highest amount of total sugar reported form *Pithecellobium dulce* where as the lowest form *Gliricidia sepium*. The same trend also observed for starch content. The highest amount of oil obtained from the seeds of *Derris indica* While *Gliricida sepium* had the lowest amount of oil. (Table-1).

Table 1: Nutritional value of leguminous seeds

Name of the species	Protein (mg/gm)	Reducing sugars (mg/gm)	Non-reducing sugar (mg/gm)	Total sugar (mg/gm)	Starch (mg/gm)	Oil (%)
Acacia nilotica	17.19	4.40	44.74	49.14	44.22	9.01
Albizia lebbeck	44.50	12.54	15.66	28.2	25.38	7.09
Bauhinia purpurea	18.15	5.14	43.90	49.04	44.13	25.32
Butea monosperma	25.58	0.80	60.34	61.14	55.02	16.68
Cassia fistula	18.3	0.80	34.00	34.8	39.52	7.95
Delonix regia	33.71	6.30	21.94	28.24	25.41	11.01
Derris indica	37.56	0.52	33.72	34.24	30.81	33.08
Gliricida sepium	50.80	0.76	10.48	11.24	10.11	4.49
L. leucocephala	40.13	0.36	25.78	26.14	23.52	6.45
P. pterocarpum	31.35	0.68	20.86	21.54	19.38	9.16
Pithecolobium dulce	33.71	5.84	59.80	65.64	59.04	16.90

Among non-leguminous seeds the highest amount of protein obtained from *Gmelina arborea* while, *Mimusops elengi* had the lowest amount. The estimated value for reducing and non-reducing sugar suggested that the highest amount of reducing sugar was obtained from *Mimusops elengi*, while the lowest obtained from *Adansonia digitata*. In case of non-reducing sugar the highest amount also obtained from the seeds of *Mimusops elengi* while *Ailanthus excelsa* had the lowest amount of non-reducing sugar. The same trend also observed for the total sugar and starch content. The highest level of oil percentage obtained from the *Gmelina arborea* and lowest percentage from *Mimusops elengi* (Table 2).

Table 2: Nutritional value of non- leguminous seeds

Name of the species	Protein (mg/ gm)	Reducing sugars (mg/gm)	Non-reducing sugar (mg/ gm)	Total sugar (mg/gm)	Starch (mg/gm)	Oil (%)
Adansonia digitata	27.91	0.10	18.34	18.44	19.34	31.44
Aegle marmalose	37.29	0.38	8.12	8.50	7.65	47.29
Ailanthus excelsa	19.87	0.88	0.86	1.74	1.56	50.73
Azadirachta indica	24.91	0.74	24.20	24.29	21.86	43.74
Gmelina arborea	58.42	1.50	1.60	3.10	2.79	51.98
Melia azedarach	26.79	0.62	9.68	10.30	9.27	38.36
Mimusops elengi	6.03	19.40	83.50	102.9	92.61	13.47
Moringa oleifera	23.86	6.64	16.56	23.20	20.86	34.69
Sapindus laurifolius	24.25	1.50	20.50	22.00	19.8	45.67
Sterculia foetida	23.62	0.32	24.92	25.25	22.75	51.65
Thespesia populnea	21.5	0.26	30.04	30.30	27.27	29.18

From the estimation for biochemical constituents of seeds, out of 22 species 17 species (9 leguminous and 8 non – leguminous): *Albizia lebbeck, Peltophorum pterocarpum, Gliricida sepium, Pithecolobium dulce, Delonix regia, Leucaena leucocephala, Derris indica, Acacia nilotica, Butea monosperma, Moringa oleifera, Aegle marmalose, Adansonia digitata, Thespesia populnea, Gmelina arborea, Sapindus laurifolius, Melia azedarach, Sterculia foetida* had fairly good amount of protein. The seeds of 16 (10 leguminous and 6 non-leguminous) species: *Albizia lebbeck, Peltophorum pterocarpum, Pithecolobium dulce, Delonix regia, Leucaena leucocephala, Bauhinia purpurea, Acacia nilotica, Butea monosperma, Moringa oleifera, Thespesia populnea, Mimusops elengi, Sapindus laurifolius, Azadirachta indica, Sterculia foetida* found rich with the good amount of total sugar. Seeds of 14 species (9 leguminous and 5 non-leguminous) species: *Albizia lebbeck, Gliricida sepium, Pithecolobium dulce, Delonix regia, Leucaena leucocephala, Derris indica, Acacia nilotica, Butea monosperma, Bauhinia purpurea, Moringa oleifera, Thespesia populnea, Mimusops elengi, Azadirachta indica, Sterculia foeitida* found enriched with starch content. In case of oil content seeds of 10 non-leguminous species and seeds of 2 leguminous species: *Derris indica, Bauhinia purpurea, Moringa oleifera, Aegle marmalose, Adansonia digitata, Thespesia populnea, Gmelina arborea, Sapindus laurifolius, Melia azedarach, Sterculia foetida, Ailanthus excelsa, Azadirachta indica* had fairly good percentage of oil.

Generally, leguminous seeds contain high level of protein but in our study, we obtained high level of protein from the non-leguminous species *Gmelina arborea* (58.42 mg/gm). In case of reducing and non-reducing sugar high level obtained from the non-leguminous species *Mimusops elengi* 19.40,mg/gm and 83.50 mg/gm respectively. The high level of total sugar also obtained with non-leguminous species *Mimusops elengi* (102.9 mg/gm). The high level of starch obtained from non-leguminous species *Mimusops elengi* (92.61 mg/gm). The high level of oil also obtained with non-leguminous species *Gmelina arborea* (51.98 %), *Sterculia foetida*

(51.65 %) and *Ailanthus excelsa* (50.73 %). Hence non-leguminous species have comparatively more nutrient contents.

The seeds of every plant species consists of different proportion of protein, carbohydrates and fats (Preiss and Levi, 1980). Seed proteins represents the major source of protein for both human and animal nutrition (Daussant, 1983). The bulk of the proteins in seeds is formed by the storage proteins and its nutrition quality largely depends upon the amino acid composition. Storage proteins in seed provide the amino acids for the growing seedling and it nourish the young seedling it is still heterotrophic.

Starch is the principal carbohydrate photosynthesized by means of solar energy. It is a mixture of glucans found mainly in the plant kingdom where it occurs as the principal food reserve polysaccharide and may be utilized during the growth of the plant. in case of uses of starch as a food, about 50 % of starch the starch produced in the world is intended for food purposes. Starches are generally considered in three categories : powered food starches, modified food starches, pregelatinised starches. It generally help to improve the functional properties of food products and provide a source of numerous oligosaccharides. The production of sweeteners (Layout, 1985) from starch by enzymes has been commercially successful for many years. Also it can be used in many ways other than as foodstuff such as in glue, coating, sizing and flocculating agents, chemicals and building materials. (Koch and Roper, 1988). The carbohydrates, proteins and fats are manufactured from the agriculture produce for food and non-food purpose. Forest, hitherto has remained commercially unexploited, although forest dwellers depend on the forest for their food at the time of scarcity and drought. Such high nutrient content of these seeds may be used as human diet. Soni (1995) suggested that forest could also be tapped for the commercial production of starch. Starch depending on the physico-chemical properties may be used in foodstuffs for improvement of their functional properties.

References

Agboola, D A and Kadiri, M (1998). Chemical analysis of seeds : Determination of cation composition, total nitrogen and crude protein in seeds of Prosopis Africana. *J. Trop. For.* 14 (2): 103 – 106.

Bonner, F T (1973). Timing collection of Samaras of Fraxinus pensylavanica Marsh in the Southern United States. Proc. IUFRO Intl. Symp. on seed processing., Bergen, Norway.

Bukhari, Y M (2002). Seed weight and protein concentration of whole seed, cotyledons and seed coat of 57 accessions of Acacia and Prosopis (Mimosaceae): Phylogeneetic and nutritional implications. Ind. For. 128(6): 686 –699.

Daussant, J and Skaknen A (1983). in Dussant. J. Mosse, J. and VAUGHAN J. (eds.) Seed Protein. Academic Press, London, 101 – 103.

Gupta, O P; Mishra, A; Mishra, P K and Agrawal, R (1995). Biochemical analysis of the seeds of some forest trees. J. Trop. For. 11(IV): 2270 – 272.

Kadam, V B (2000a). Nutritional status of seeds of tree species of South Gujarat Forests. J. Adv. In Sci. and Tech. 3(1 & 2): 77 – 82.

Kadam, V B (2000b). Protein and amino acids in seeds of some tree species of forests of South Gujarat. Geobios, 27: 117 –119.

Kadam, V B (2001). Protein and amino acid in seeds of some medicinally important tree species. Plant Archives, 1 (1 & 2): 57 – 59.

Kadam, V B; Krishnamurthy, R and Parabia, M H (1996). Nutritional status of seeds of some tree species. J. Env. Bio., 5 : 122 – 123.

Koch, H and Roper H (1988). New industrial Products from starch, Starch/Storkei, 40 : 121 –131.

Layout, J J M (1985). Conversion of Liquified Starch into Glucose Using a Novel Glucoamylase system. Starch/Storkei, 37 : 157 - 161.

Lowry, O H; Rosebrough, N J; Farr, A L and Randall, R J (1951). Protein measurement with folin-phenol reagent ., J. Bio. Che. 193: 269 – 275.

Mahadevan, A and Sridhar, R (1974). Methods in Physiological Plant Pathology (4th edition). Sivakami Publication, Chennai. pp – 124.

Nelson, N (1944). A Photometric adaption of the Somogyi method for the determination of glucose., J. Bio. Chem. 152: 375.

Singh, Lal (1987). Practical Agriculture Chemistry and Soil Science. Shiva Printers, Dehra Dun for Bishen Singh and Mehendra Pal Singh. pp: 160 –161.

Soni, P L (1995). Food from forest. Ind. For. 121 (9) : 837 – 8845.

Tomar, A K; Khan, M A; Tomar, G S and Kumar, S (1996). Forest leguminous trees – a source of human nutrition. J. Trop. For. 12 (IV): 189 – 193.

CHAPTER 4

ENVIRONMENTAL IMPACT ASSESSMENT FOR A BETTER TOMORROW

Dr. Cyril Daniel, Dr. Sushma Tiwari

Department. Of English, Govt. Arts and Commerce Girls College, Raipur, C.G., India

Abstract

Environmental Impact Assessment (EIA) is a vital tool for achieving sustainable development by evaluating the potential environmental consequences of proposed projects before they are implemented. This chapter examines the role of EIA in fostering a sustainable future, discussing its methodological framework, including screening, scoping, baseline study, impact prediction, mitigation measures, public participation, and reporting. The results highlight the effectiveness of EIA in preventing environmental degradation, promoting sustainable practices, and enhancing public participation in decision-making processes. However, challenges such as variability in EIA quality, the need for better mitigation implementation, and the integration of climate change considerations are also discussed. By addressing these challenges, EIA can continue to play a crucial role in safeguarding the environment and ensuring that development projects contribute positively to environmental sustainability.

Keywords: Environmental Impact Assessment (EIA), Sustainable Development, Mitigation Measures, Public Participation, Climate Change Integration.

Introduction

Environmental Impact Assessment (EIA) has emerged as a critical tool for sustainable development, offering a structured framework for evaluating the potential environmental consequences of proposed projects or actions. With the global push toward sustainability and the need to mitigate adverse environmental impacts, EIA serves as a decision-making tool that integrates environmental considerations into the planning and development processes (Glasson, Therivel, & Chadwick, 2013). This chapter explores the significance of EIA in fostering a better tomorrow by ensuring that environmental consequences are identified, evaluated, and mitigated before the implementation of any major project or policy.

Methodological Framework

The methodology of EIA is a systematic process designed to assess the environmental effects of a proposed project or action before any decision is made to move forward. This section outlines the general steps involved in conducting an EIA, which typically includes screening, scoping, baseline study, impact prediction, mitigation measures, public participation, and the final EIA report (Morris & Therivel, 2001).

- **Screening:** The first step involves determining whether a project requires an EIA and the level of assessment needed. This decision is often based on the size, nature, and potential environmental impact of the project (Glasson & Salvador, 2000). Projects with significant environmental impacts are subjected to a full EIA, while smaller projects may only require a preliminary assessment (Wood, 2003).
- **Scoping:** In this stage, the key issues and potential impacts of the project are identified. The scoping process defines the scope of the EIA, including the environmental factors to be considered, the geographical area to be covered, and the time frame for the assessment (Caldwell, 1998). This stage is crucial for focusing the EIA on the most significant environmental concerns (Weston, 2000).
- **Baseline Study:** A baseline study involves collecting data on the existing environmental conditions in the project area. This data serves as a benchmark against which the potential impacts of the project can be measured (Sadler, 1996). Baseline studies typically cover various environmental aspects, including air and water quality, biodiversity, soil conditions, and socio-economic factors (Noble, 2000).
- **Impact Prediction and Evaluation:** This step involves predicting the potential environmental impacts of the proposed project, both positive and negative. Various tools and models are used to forecast these impacts, such as Geographic Information Systems (GIS), remote sensing, and simulation models (Fischer, 2007). The predicted impacts are then evaluated in terms of their significance, likelihood, and reversibility (Morgan, 2012).
- **Mitigation Measures:** Once the potential impacts have been identified and evaluated, mitigation measures are proposed to avoid, reduce, or compensate for any adverse effects. These measures can include changes to the project design, the implementation of environmental management plans, or the restoration of affected areas (Lawrence, 2003).

- **Public Participation**: Public involvement is a critical component of the EIA process. It ensures that the concerns and values of local communities and other stakeholders are considered in the decision-making process (Kørnøv & Thissen, 2000). Public participation can take various forms, including public hearings, workshops, and consultations (Sadler, 1996).
- **EIA Report:** The final step in the EIA process is the preparation of the EIA report, which presents the findings of the assessment, including the predicted impacts, proposed mitigation measures, and any residual effects. The report also includes an environmental management plan outlining the strategies for monitoring and managing the environmental impacts during the implementation of the project (Lawrence, 2003).

Results and Discussion

The results of an EIA are critical for informing decision-makers about the potential environmental impacts of a proposed project. In this section, we discuss the outcomes of EIA studies, focusing on how they contribute to sustainable development and environmental protection.

- **Environmental Protection:** EIA has been instrumental in preventing significant environmental degradation by identifying potential impacts early in the planning process. For instance, EIAs have led to the modification or abandonment of projects that posed unacceptable risks to sensitive ecosystems, such as wetlands, forests, and coral reefs (Pope, Annandale, & Morrison-Saunders, 2004). The implementation of mitigation measures has also helped minimize the adverse effects of projects on air and water quality, biodiversity, and local communities (Cashmore, 2004).
- **Sustainable Development:** By integrating environmental considerations into the decision-making process, EIA contributes to the achievement of sustainable development goals (Bond & Morrison-Saunders, 2009). It ensures that economic development does not come at the expense of the environment and that natural resources are used sustainably. EIA also promotes the adoption of cleaner technologies and practices, which can reduce the environmental footprint of development projects (Glasson, Therivel, & Chadwick, 2013).
- **Public Participation and Social Acceptance:** One of the key strengths of the EIA process is its emphasis on public participation (Sadler, 1996). By involving local communities and other stakeholders in the assessment process, EIA fosters transparency and accountability. This, in turn, can enhance the social acceptance of projects, as stakeholders feel that their concerns have been considered and addressed (Glasson & Salvador, 2000). Public participation also helps identify potential social impacts that may not have been apparent during the initial stages of the assessment (Noble, 2000).

- **Challenges and Limitations:** Despite its successes, the EIA process is not without its challenges. One of the main limitations is the variability in the quality and rigor of EIA studies, which can result in incomplete or biased assessments (Morgan, 2012). This can be due to a lack of technical expertise, inadequate data, or insufficient resources. Additionally, the effectiveness of mitigation measures is often uncertain, as they may not be properly implemented or monitored (Lawrence, 2003). There is also a need for greater integration of climate change considerations into the EIA process, as many EIAs currently do not adequately address the long-term impacts of climate change on projects and their surrounding environments (Bond & Morrison-Saunders, 2009).

Conclusion

Environmental Impact Assessment is a vital tool for promoting environmental sustainability and safeguarding natural resources for future generations. By systematically evaluating the potential impacts of proposed projects, EIA ensures that environmental considerations are integrated into the decision-making process. The successful implementation of EIA has led to the protection of sensitive ecosystems, the promotion of sustainable development, and the enhancement of social acceptance for development projects (Glasson, Therivel, & Chadwick, 2013). However, there is still room for improvement in the EIA process. To better address the environmental challenges of tomorrow, it is essential to enhance the quality and rigor of EIA studies, improve the implementation and monitoring of mitigation measures, and integrate climate change considerations more effectively (Fischer, 2007). By doing so, EIA can continue to play a crucial role in achieving a sustainable and environmentally responsible future (Sadler, 1996).

References

Bond, A., & Morrison-Saunders, A. (2009). Sustainability Appraisal: Jack of All Trades, Master of None? Impact Assessment and Project Appraisal, 27(4), 321-329.

Caldwell, L. K. (1998). Implementing Policy through Procedure: Impact Assessment and the National Environmental Policy Act. Environmental Professional, 20, 25-42.

Cashmore, M. (2004). The Role of Science in Environmental Impact Assessment: Process and Procedure versus Purpose in the Development of Theory. Environmental Impact Assessment Review, 24(4), 403-426.

Fischer, T. B. (2007). Theory and Practice of Strategic Environmental Assessment. Earthscan.

Glasson, J., & Salvador, N. (2000). EIA in Brazil: A Procedures–Practice Gap. Environmental Impact Assessment Review, 20(2), 191-204.

Glasson, J., Therivel, R., & Chadwick, A. (2013). Introduction to Environmental Impact Assessment. Routledge.

Kørnøv, L., & Thissen, W. A. H. (2000). Rationality in Decision- and Policy-Making: Implications for Strategic Environmental Assessment. Impact Assessment and Project Appraisal, 18(3), 191-200.

Lawrence, D. P. (2003). Environmental Impact Assessment: Practical Solutions to Recurrent Problems. John Wiley & Sons.

Morgan, R. K. (2012). Environmental Impact Assessment: The State of the Art. Impact Assessment and Project Appraisal, 30(1), 5-14.

Morris, P., & Therivel, R. (2001). Methods of Environmental Impact Assessment. Spon Press.

Noble, B. F. (2000). Strategic Environmental Assessment: What Is It? & What Makes It Strategic? Journal of Environmental Assessment Policy and Management, 2(2), 203-224.

Pope, J., Annandale, D., & Morrison-Saunders, A. (2004). Conceptualising Sustainability Assessment. Environmental Impact Assessment Review, 24(6), 595-616.

Sadler, B. (1996). International Study of the Effectiveness of Environmental Assessment. Final Report, Canadian Environmental Assessment Agency, Ottawa, Canada.

Weston, J. (2000). EIA, Decision-Making Theory and Screening and Scoping in UK Practice. Journal of Environmental Planning and Management, 43(2), 185-203.

Wood, C. (2003). Environmental Impact Assessment: A Comparative Review. Prentice Hall.

CHAPTER 5

HARNESSING INDIGENOUS KNOWLEDGE FOR CLIMATE ADAPTATION: INTEGRATING TRADITIONAL PRACTICES WITH MODERN SCIENCE

Dr. Sweta Gaikwad

Assistant Professor (Botany), Dept. of Life-Sciences,
Shri Shankaracharya Professional University,
Bhilai, Chhattisgarh, India
e-mail: swetagaikwad7@gmail.com

Abstract

Communities around the world face serious issues as a result of climate change, necessitating innovative and effective adaptation strategies. This paper explores the potential of harnessing indigenous knowledge related to climate adaptation, emphasizing the integration of traditional practices along with modern scientific approaches. Indigenous communities have a wealth of environmental knowledge, developed over centuries, that gives valuable insights into the sustainable land management, biodiversity conservation, and climate resilience. By examining case studies from various regions, we highlight successful instances where indigenous knowledge has complemented scientific methods, resulting in enhanced adaptive capacity and ecosystem health. This integration not only fosters more holistic and culturally relevant adaptation strategies but also promotes the inclusion and empowerment of indigenous people in climate governance. The paper argues for the necessity of bridging the gap between traditional and scientific knowledge systems to address the multifaceted impacts of climate change effectively. Ultimately, the synthesis of these knowledge systems may lead to more resilient communities and ecosystems, providing a robust framework for global climate adaptation efforts.

Keywords: Indigenous Knowledge, Traditional Practices, Climate Resilience, Climate Governance, Ecosystem Health.

Introduction

One of the most important issues of the twenty-first century is climate change, which has an impact on communities, economies, and ecosystems all around the world. As the frequency and intensity of climate-related events increase, the need for effective adaptation strategies becomes ever more critical. Traditional approaches to climate adaptation have predominantly relied on modern scientific methods and technological advancements. However, there is a growing recognition of the value that indigenous knowledge can bring to these efforts. Indigenous knowledge is made up of the concepts, abilities, and ideologies that have been evolved over many generations by communities that have interacted with their natural environment. These knowledge systems are rich in ecological insights and sustainable practices, honed over generations through careful observation and adaptation to changing environments. Indigenous communities around the world have successfully managed and conserved their resources in ways that are often more sustainable and adaptable than contemporary methods. (The most recent report from the Intergovernmental Panel on Climate Change (IPCC) states that indigenous and local knowledge can influence how people perceive and understand the risk of climate change as well as offer reliable and practical observations, information, and solutions (Schipper et al., 2022).

Integrating indigenous knowledge with modern science offers a promising pathway to enhance climate adaptation strategies. This approach not only bridges the gap between scientific and traditional knowledge systems but also ensures that adaptation measures are culturally relevant and socially inclusive. By utilizing the advantages of both knowledge systems, we may create more robust and all-encompassing responses to the effects of climate change. Understanding the significance and efficacy of indigenous knowledge systems has grown as climate change has gained prominence on a global scale. This is particularly true of how indigenous knowledge and practices may be used to support successful adaptation and mitigation efforts. (Zougmore et al., 2023).

This chapter explores the potential of integrating indigenous knowledge into modern climate adaptation practices. Through a series of case studies and examples, we illustrate how traditional practices can complement scientific approaches, leading to improved outcomes for both people and ecosystems. We also discuss the opportunities and challenges associated with this integration, emphasizing the importance of respecting and valuing indigenous contributions to climate resilience. In conclusion, harnessing indigenous knowledge for climate adaptation is not only a matter of environmental necessity but also a step towards greater social justice and equity. By recognizing and incorporating the wisdom of indigenous peoples, we can create more effective, inclusive, and sustainable solutions to the global challenge of climate change. A variety of cultural practices, traditions, wisdom, and worldviews might be characterized as such knowledge (Taylor et al., 2023).

1. Understanding Indigenous Knowledge Systems

Indigenous Knowledge Systems (IKS) may refer to the complex array of knowledge, beliefs, practices, and skills that Indigenous peoples have developed and honed over millennia through direct interaction with their natural environments. An emphasis on IKS suggests the archeology and re-appropriation of information that was not permitted to "be," in order to advance human understanding and develop, preserve, and disseminate it (Odora Hoppers, 2021). These systems are holistic, encompassing various domains such as agriculture, medicine, astronomy, ecology, and spirituality. They are often transmitted orally through generations and are deeply embedded in the cultural and social practices of Indigenous communities.

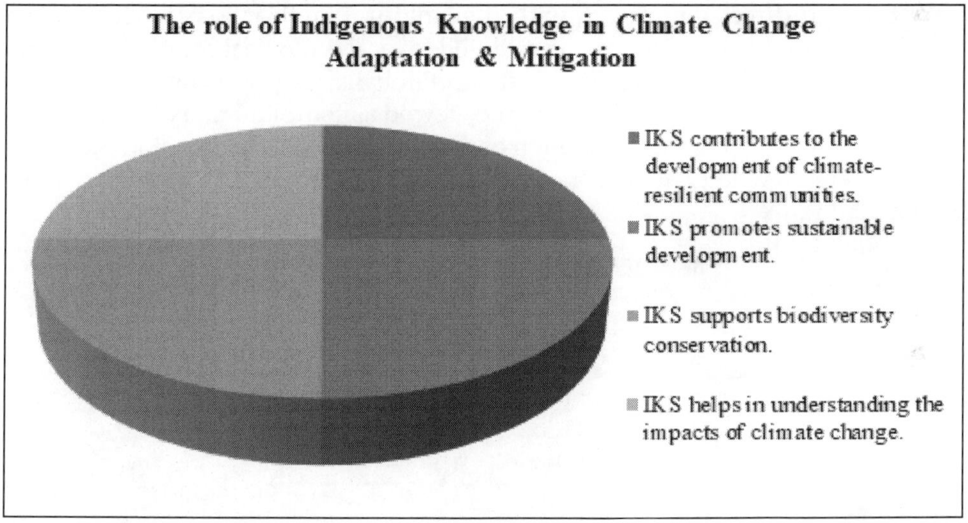

Fig 1: Role of Indigenous Knowledge System in Climate Change Adaptation and Mitigation

Key Characteristics of Indigenous Knowledge Systems

a. **Holistic and Integrated**: IKS are characterized by their holistic nature, integrating spiritual, social, and ecological dimensions. Knowledge is not compartmentalized but is seen as interconnected and interdependent.

b. **Context-Specific and Adaptable**: Indigenous knowledge is deeply rooted in specific geographical and cultural contexts, making it highly adapted to local environments and conditions. It is dynamic and evolves in response to changes in the environment and society.

c. **Sustainable Practices**: Many Indigenous practices are inherently sustainable, emphasizing balance and harmony with nature. These practices often include sophisticated resource management strategies that ensure the long-term health of ecosystems.

d. **Community-Based and Participatory**: IKS are typically shared and maintained through community participation. Knowledge is collectively owned and its transmission involves communal activities and rituals.

e. **Oral Tradition and Storytelling**: Oral tradition is a primary mode of knowledge transmission in many Indigenous cultures. Stories, songs, and rituals play a crucial role in preserving and conveying knowledge across generations.

Indigenous knowledge provides comparatively rich and diverse understandings, practices, and worldviews developed by indigenous people over thousands of years through direct interaction with their natural environments. This body of knowledge is highly specialized and adapted to local ecological conditions, making it invaluable for environmental stewardship. Globally, there is growing recognition that adaptation and mitigation ought to be viewed as complimentary to one another. Nonetheless, enhancing the integration of climate change mitigation and adaptation methods is not a wholly novel concept in India, particularly in northwestern India (Kumar, 2014).

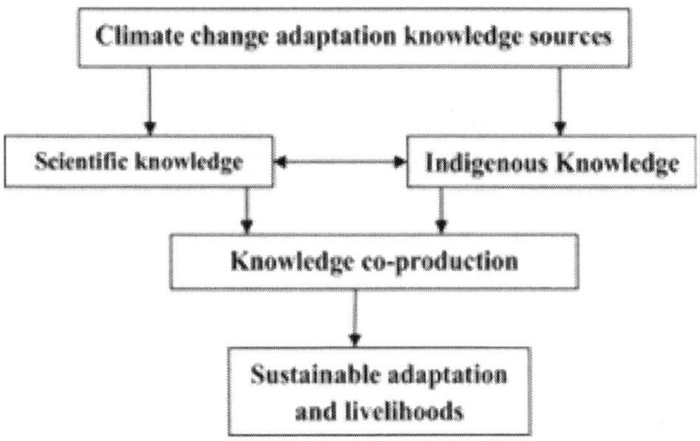

Fig 2: Flow-chart for sources of climate change adaptation knowledge in climate change adaptation (Mekonnen et al., 2021)

2. Traditional Practices for Climate Adaptation

Over the years, traditional knowledge—the wisdom, knowledge, and traditions of indigenous people that have been acquired through experience and verbally transmitted from generation to generation—has been crucial in helping to solve a number of issues, including those pertaining to climate change and variability. Native Americans who reside near natural resources frequently watch what goes on around them and are the first to notice and adjust to any change (Gyampoh et al., 2009).

Indigenous communities worldwide have developed a variety of traditional practices to adapt to and mitigate the impacts of climate variability and change. There are gaps in both scientific and traditional knowledge, therefore it would be better to combine the two to create an additional influence that can support sustainable development, create livelihoods, and increase landscape resilience during climate change (Mekonnen, 2021). These practices, honed over centuries, offer sustainable and effective strategies for dealing with climate-related challenges. Here, we explore key traditional practices across different domains.

Agricultural Practices

1. Crop Diversity and Rotation:

- **Poly-cropping:** Planting multiple crops simultaneously to enhance biodiversity, improve soil health, and reduce pest outbreaks.
- **Crop Rotation:** Alternating crops in a specific sequence to maintain soil fertility and control pests and diseases.

2. Traditional Irrigation Techniques:

- **Floodplain Irrigation:** Utilizing natural flood cycles to irrigate crops, recharge groundwater, and deposit nutrient-rich silt.
- **Terracing:** Constructing terraces on hillsides to reduce soil erosion, retain water, and create flat areas for agriculture.

3. Indigenous Crop Varieties:

- **Drought-Resistant Varieties:** Selecting and cultivating crop varieties that are naturally resistant to drought and other extreme weather conditions.
- **Seed Saving and Sharing:** Preserving and exchanging seeds of locally adapted crops to maintain genetic diversity and resilience.

Water Management

1. Rainwater Harvesting:

- **Traditional Cisterns and Tanks:** Building structures to collect and store rainwater for use during dry periods.
- **Check Dams and Ponds:** Constructing small ponds and dams to capture and then store runoff water for agriculture and domestic use.

2. Sacred Water Bodies:

- **Protection of Sacred Groves and Springs:** Managing water resources through cultural and spiritual practices that ensure their protection and sustainable use.
- **Community Water Management:** Collective management of water resources, including shared wells and irrigation systems.

3. Wetland and Floodplain Management:

- **Wetland Restoration:** Preserving and restoring wetlands to enhance their natural flood control and water purification functions.
- **Seasonal Flooding:** Utilizing seasonal flooding for soil nutrient replenishment and crop irrigation.

Forestry and Land Use

1. Agroforestry:

- **Tree Integration:** Combining trees and shrubs along with crops and livestock to create more productive, diverse, and sustainable land-use systems.
- **Alley Cropping:** Planting rows of shrubs or trees between crops to reduce wind erosion, improve soil fertility, and provide additional resources such as fruit or fodder.

2. Fire Management:

- **Controlled Burns:** Using prescribed burns to manage forest health, reduce fuel loads, and prevent larger, uncontrolled wildfires.
- **Cultural Burning:** Indigenous fire practices that promote biodiversity and ecosystem resilience.

3. Land Stewardship:

- **Rotational Grazing:** Managing livestock grazing patterns to prevent overgrazing, promote vegetation regrowth, and maintain soil health.
- **Sacred Sites Protection:** Preserving culturally significant lands that often serve as biodiversity hotspots and ecological refuges.

Disaster Risk Reduction

1. Early Warning Systems:

- **Environmental Indicators:** Using natural signs, such as animal behavior and plant phenology, to predict and prepare for weather events like floods, droughts, and storms.
- **Traditional Knowledge Sharing:** Passing down knowledge through storytelling and community gatherings to inform and prepare future generations.

2. Resilient Infrastructure: Traditional Housing Designs involves constructing homes using local materials and techniques that enhance resilience to extreme weather, such as elevated houses in flood-prone areas

3. Modern Scientific Approaches to Climate Adaptation:

Modern scientific approaches to climate adaptation leverage technological innovations, data-driven solutions, and robust policy frameworks to address the multifaceted challenges posed by climate change. Though it has trailed behind, the creation of a solid knowledge base to assist adaptation globally is advised to speed up progress. Both the existing and newly proposed programs have a significant emphasis on practice-oriented research that involves active stakeholder participation (Swart, 2014). By integrating these approaches with traditional knowledge and practices, we can develop comprehensive, sustainable, and effective strategies to build resilient communities and ecosystems.

Technological Innovations

1. Crop Diversity and Rotation:

- **Poly-cropping:** Planting multiple crops together to enhance biodiversity, improve soil health, and reduce pest outbreaks.
- **Crop Rotation:** crop rotation following a set schedule to preserve soil fertility and manage pests and illnesses.

2. Traditional Irrigation Techniques:

- **Floodplain Irrigation:** Utilizing natural flood cycles to irrigate crops, recharge groundwater, and deposit nutrient-rich silt.
- **Terracing:** Constructing terraces on hillsides to reduce soil erosion, retain water, and create flat areas for agriculture.

3. Wetland and Floodplain Management:

- **Wetland Restoration:** Preserving and restoring wetlands to enhance their natural flood control and water purification functions.
- **Seasonal Flooding:** Utilizing seasonal flooding for soil nutrient replenishment and crop irrigation.

Data-Driven Solutions

1. Climate Modeling and Forecasting: Using sophisticated models to predict climate patterns and potential impacts, provides critical information for planning and decision-making, helping communities prepare for and mitigate climate risks.

2. Geographic Information Systems (GIS): Mapping and analyzing spatial data to understand and manage land use, natural resources, and environmental changes, supports effective planning and resource management, aiding in the design of adaptive strategies.

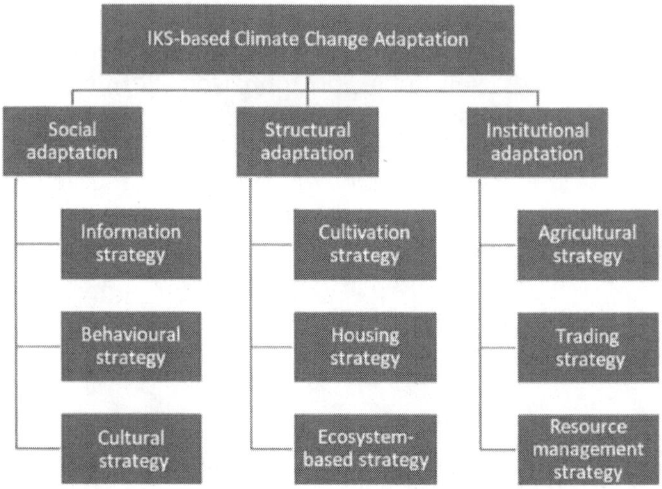

Fig 3: Indigenous Knowledge System based climate change adaptation and its types (Mbah et al., 2021).

3. Big Data Analytics: Collecting and analyzing large datasets from various sources (e.g., satellite imagery, weather stations, social media) to identify trends and patterns, enhances the understanding of climate impacts and helps in developing targeted adaptation measures.

4. Community-Based Adaptation: Giving local communities the tools they need to create and carry out their own adaptation plans, ensures that adaptation measures are locally relevant and culturally appropriate, enhancing community resilience and ownership.

4. Synergies Between Traditional Practices and Modern Science on Climate Adaptation

In this sense, we believe that indigenous knowledge is just as important as scientific knowledge and that the two must be integrated through multiple evidence base approach for climate change adaptation and mitigation. Knowledge systems rarely develop in isolation as they normally tend to cross-fertilize and benefit from each other (Makondo and Thomas, 2018). The integration of traditional practices with modern scientific approaches offers a powerful synergy for climate adaptation, combining the depth of local, time-tested knowledge with the breadth of advanced, evidence-based methodologies. Traditional practices, such as crop diversity, rainwater harvesting, and controlled burning, are deeply rooted in an understanding of local ecosystems and have proven resilient over centuries. Modern science, with its technological innovations and data-driven solutions, can enhance these practices by providing precise tools for monitoring, prediction, and optimization. For instance, precision agriculture technologies can improve traditional crop rotation methods, optimizing planting schedules and resource use. Geographic Information Systems (GIS) can map and protect sacred groves, integrating cultural preservation with environmental management. Moreover, community-based adaptation plans can be enriched by both indigenous knowledge and scientific research, ensuring that adaptation strategies are both culturally relevant and scientifically robust. This collaboration not only increases the effectiveness of adaptation measures but also promotes sustainability, resilience, and inclusivity in addressing climate challenges.

5. Challenges and Barriers to Integration of Traditional and Modern Methods for Climate Adaptation

Integrating traditional practices with modern scientific methods for climate adaptation faces several challenges and barriers. One major obstacle is the knowledge gap and lack of mutual understanding between indigenous communities and scientific researchers. Traditional knowledge is often passed down orally and may not be documented in a way that is accessible or comprehensible to scientists. Conversely, modern scientific methods and terminologies can be complex and unfamiliar to local communities. Additionally, there are cultural sensitivities and trust issues, as indigenous peoples may be wary of external interventions that have historically marginalized their practices and perspectives. Institutional and policy hurdles also exist, as frameworks for climate adaptation may not recognize or support the integration of indigenous knowledge, resulting in a dearth of resources and money for these kinds of projects. Considerable progress has been made in a number of industries' policies and practices to adapt to climate change and the hazards it poses (Akturk, 2022). Furthermore, the scalability of traditional practices can be a challenge when addressing widespread climate impacts, requiring innovative approaches to adapt these methods to larger contexts. Adaptation to climate change requires "adjustments, or changes in decision environments, which might ultimately enhance resilience or reduce vulnerability to observed or expected changes in climate" (Adger, 2007). Barriers hampering their adaptive capacity were identified as: a lack of access to information, a lack of access to extension services, limited awareness and knowledge, and limited financial options (Antwi-Agyei et al., 2015). Overcoming these barriers requires fostering mutual respect, collaboration, and co-creation of knowledge between indigenous communities and scientists, as well as developing inclusive policies that value and incorporate diverse knowledge systems.

6. Strategies for Effective Integration of Traditional and Modern Methods for Climate Adaptation

Nowadays, it is crucial for decision-makers at all levels to take into account how climate change affects various species, disturbances, and environmental circumstances (Mac Dicken et al., 2015). Effective integration of traditional and modern methods for climate adaptation requires a multifaceted approach. Through community-based adaptation, the local population's resistance to climatic unpredictability and change can be strengthened as well as reduced. Understanding community-level reactions is essential for creating effective adaptation programs in the face of growing climate-related concerns (Suhaeb et al., 2024). Over the years, subsistence farmers have always adapted their farming practices to some of these changes. Therefore, adopting traditional methods will greatly aid individuals of rural communities in their adaptation to climate change (Mugambiwa, 2018). Engaging and empowering local communities in collaborative decision-making and participatory research ensures that their knowledge and needs are incorporated into adaptation strategies. Documenting traditional practices through oral histories and digital archives preserves this valuable knowledge, while creating knowledge exchange platforms facilitates sharing between communities and researchers. Policy support is crucial, with inclusive policies and dedicated funding fostering integration efforts. Educational programs and awareness campaigns promote mutual understanding and respect for both traditional and scientific approaches. Technological innovations, such as GIS for managing sacred groves and drones for precision agriculture, can enhance traditional practices. Establishing innovation hubs encourages collaboration between indigenous practitioners and scientists. Finally, participatory monitoring and adaptive management ensure that strategies remain effective and culturally relevant, promoting resilience, sustainability, cultural preservation, and community empowerment.

7. Policy Recommendations

Policy recommendations for integrating traditional and modern methods for climate adaptation should focus on fostering inclusive, collaborative, and sustainable practices. Recent proposals for more robust policy integration in the context of climate change adaptation have arisen to guarantee prompt, sufficient, and efficient measures (Biesbroek, 2021). Governments and institutions should develop policies that recognize and legitimize traditional knowledge systems, ensuring they are valued alongside scientific approaches. This can be achieved by creating legal frameworks that protect indigenous intellectual property rights and promote equitable benefit-sharing. Policies should allocate funding and resources specifically for projects that integrate traditional practices with modern science, encouraging community-led initiatives and capacity building. Establishing advisory bodies comprising indigenous leaders, scientists, and policymakers can facilitate dialogue and co-creation of adaptation strategies. Education policies should incorporate traditional ecological knowledge into curricula at all levels, fostering a culture of respect and understanding. The "responsiveness trap" and the accumulation of policies have made the goal of climate policy integration even more popular in recent years (Adam et al., 2019). Governments are frequently asked by citizens to take greater action on climate change, and in response, they often produce more policy outputs. This leads to a complicated and sometimes contradictory policy environment, which demands for tighter integration to mitigate its negative impacts. Furthermore, international cooperation is essential, with agreements and frameworks supporting cross-cultural knowledge exchange and joint research initiatives. By implementing these policy recommendations, we can create a more resilient, adaptive, and inclusive approach to climate change that harnesses the strengths of both modern and traditional methods (Knill et al., 2021).

Conclusion

Harnessing Indigenous knowledge for climate adaptation by integrating traditional practices with modern science presents a powerful and holistic approach to address the multifaceted challenges of climate change. Indigenous knowledge systems, developed over centuries, offer sustainable and context-specific solutions that enhance community resilience and environmental stewardship. When combined with the precision, technological advancements, and data-driven insights of modern science, these practices can lead to more effective and comprehensive climate adaptation strategies. The integration of these knowledge systems requires mutual respect, inclusive policies, and collaborative efforts that recognize and value the contributions of indigenous communities. By fostering such synergies, we can create adaptation measures that are not only scientifically robust but also culturally appropriate and sustainable. This approach not only helps mitigate the impacts of climate change but also promotes biodiversity, supports local economies, and preserves cultural heritage. Ultimately, the collaboration between traditional and modern methods holds the potential to build more resilient and adaptive communities worldwide, ensuring a sustainable future for generations to come.

References

Adam, C., Hurka, S., Knill, C., & Steinebach, Y. (2019). Policy accumulation and the democratic responsiveness trap. Cambridge University Press.

Adger, W. N., Agrawala, S., Mirza, M. M. Q., Conde, C., O'Brien, K., Pulhin, J., & Takahashi, K. (2007). Assessment of adaptation practices, options, constraints and capacity. ML Parry, OF Canziani, JP Palutikof, PJ van der Linden and CE Hanson, (Eds.) Climate Change 2007: Impacts, Adaptation and Vulnerability. Contribution of Working Group II to the Fourth Assessment Report of the Intergovernmental Panel on Climate Change.

Aktürk, G. (2022). A systematic overview of the barriers to building climate adaptation of cultural and natural heritage sites in polar regions. Environmental Science & Policy, 136, 19-32.

Antwi-Agyei, P., Dougill, A. J., & Stringer, L. C. (2015). Barriers to climate change adaptation: evidence from northeast Ghana in the context of a systematic literature review. Climate and Development, 7(4), 297-309.

Biesbroek, R. (2021). Policy integration and climate change adaptation. Current Opinion in Environmental Sustainability, 52, 75-81.

Gyampoh, B. A., Amisah, S., Idinoba, M., & Nkem, J. (2009). Using traditional knowledge to cope with climate change in rural Ghana. Unasylva, *60(281/232), 70-74.*

Knill, C., Steinbacher, C., & Steinebach, Y. (2021). Balancing trade offs between policy responsiveness and effectiveness: The impact of vertical policy process integration on policy accumulation. Public Administration Review, 81(1), 157-160.

Kumar, V. (2014). Role of indigenous knowledge in climate change adaptation strategies: A study with special reference to North-Western India. J Geogr Nat Disast, 5(131), 2167-0587.

Mbah, M., Ajaps, S., & Molthan-Hill, P. (2021). A systematic review of the deployment of indigenous knowledge systems towards climate change adaptation in developing world contexts: implications for climate change education. Sustainability, 13(9), 4811.

MacDicken, K. G., Sola, P., Hall, J. E., Sabogal, C., Tadoum, M., & de Wasseige, C. (2015). Global progress toward sustainable forest management. Forest Ecology and Management, 352, 47-56.

Makondo, C. C., & Thomas, D. S. (2018). Climate change adaptation: Linking indigenous knowledge with western science for effective adaptation. Environmental science & policy, 88, 83-91.

Mekonnen, Z., Kidemu, M., Abebe, H., Semere, M., Gebreyesus, M., Worku, A., & Chernet, A. (2021). Traditional knowledge and institutions for sustainable climate change adaptation in Ethiopia. Current research in Environmental sustainability, 3, 100080.

Odora Hoppers, C. (2021). Research on Indigenous knowledge systems: The search for cognitive justice. International Journal of Lifelong Education, 40(4), 310-327.

Schipper, E. L. F., Revi, A., Preston, B. L., Carr, E. R., Eriksen, S. E., Fernandez-Carril, L. R., ... & Tebboth, M. (2022). Climate resilient development pathways.

Suhaeb, F. W., Tamrin, S., Jumadi, J., & Irwan, I. (2024). Community Adaptation Strategies to Climate Change: Towards Sustainable Social Development. Migration Letters, 21(S2), 943-953.

Swart, R., Biesbroek, R., & Lourenço, T. C. (2014). Science of adaptation to climate change and science for adaptation. Frontiers in environmental science, 2, 29.

Taylor, J. E., Poleacovschi, C., & Perez, M. A. (2023). Climate change adaptation trends among Indigenous peoples: a systematic review of the empirical research focus over the last 2 decades. Mitigation and Adaptation Strategies for Global Change, 28(6), 29.

Zougmoré, R., Segnon, A. C., & Thornton, P. (2023). Harnessing indigenous knowledge and practices for effective adaptation in the Sahel. Current Opinion in Environmental Sustainability, 65, 101389.

CHAPTER 6

SUSTAINABLE USE OF AGRICULTURAL LAND IN HIGHLY POPULATED COUNTRY

Dr. Razia Sultana[1], Dr. Sangeeta Jha Aditya[2]

[1]*Shri Kuleshwar Mahadev Government College Gobra Nawapara Raipur CG.*
raziasultana83ster@gmail.com ; https://orcid.org/0000-0002-1556-2140
[2]*Assistant Professor (Economics), Govt. Arts and Commerce Girls' College, Raipur, Chhattisgarh*
sangaditya222@gmail.com

Abstract

In order to meet the demand for food and other essentials, sustainable agriculture—farming based on environmental services—is crucial, particularly for emerging nations with dense populations. Numerous techniques exist to enhance the sustainability of agriculture, such as crop rotation, multiple cropping, mixed farming, and agroforestry. Modern traditional agriculture practices have a severe influence on the environment due to processes including deforestation, water pollution, land degradation, and water scarcity. They also demand a lot of energy, water, chemicals, and other subsidies. We will talk about modern agricultural technologies for sustainable farming practices in this chapter. These environmentally friendly techniques enable crop production without endangering people or the environment.

Keywords: Agriculture, Sustainability, Precision farming.

Introduction

In countries with high food consumption and dense populations but little land, sustainable agriculture methods are crucial. These techniques aim to produce more food, lessen environmental harm, and guarantee long-term sustainability. Over

298.3 million hectares make up the whole area of India, the second most populated country in the world, of which roughly 165.7 million hectares or nearly 55.5% of the total geographical area are used for agriculture. The net sown area is around 147.9 million hectares. This land is used for agricultural production, livestock grazing, and aquaculture. This chapter will address the key strategies for using agricultural land sustainably in highly populated countries such as India.

Sustainable Intensification

It involves all the methods to increase the efficiency of agricultural resources and reduce negative environmental and social impacts (Petersen & Snapp 2015). Netherlands, United States, china, Japan, and Canada are some of the leading countries in adaptation of advance techniques for agriculture. It involves the following practices

Vertical farming: "Vertical farming" is a type of agriculture where plants are grown in stacked layers, usually inside or in controlled environments. China and Singapore are two of the top nations for vertical forming. Plants can grow on several levels in multistory buildings, increasing yield per unit area. Precise control over temperature, humidity, light, and CO2 levels allows for optimal plant growth while lowering the danger of pests and illnesses. In addition, artificial lighting is necessary for vertical farming in order to grow crops indoors in stacked layers for maximum yield per square foot. It encourages robust plant development and quickens the germination of seeds. It enables farmers to grow crops outside of their regular growing seasons, expanding the market and maybe boosting revenue. It can produce higher yields per unit area compared to traditional agriculture as well as minimize the need for pesticides, fertilizers, and transportation. Vertical farming can be adopted in urban areas, providing fresh, locally grown food to city dwellers.

Hydroponics and aeroponics: These are creative techniques for soilless plant growth. These methods entail growing plants in nutrient-rich mists or aqueous solutions, respectively. Compared to conventional soil-based agriculture, they provide a number of benefits, particularly in areas with a shortage of water, restricted acreage, or poor soil. There are various types of hydroponics systems, including as wick systems, ebb and flow, deep water culture, and nutrient film technique (NFT). It is possible to cultivate plants in small places, with less water used, faster growth rates, and precise control over nutrients. Similar to this, aeroponics is a type of nutrient-rich water mist farming where plants are cultivated. Rapid growth and development are encouraged by the roots' continual exposure to the nutrient mist. The method is beneficial because it reduces water usage, increases oxygen availability to the roots, and grows plants in a fully sterile environment. Aeroponics is a more sophisticated technique than hydroponics, though. Many crops, such as strawberries, tomatoes, and leafy greens, are grown using these techniques. By growing food in cities, they can lessen the need for long-distance transportation.

Precision agriculture: It uses satellite technology to accurately manage the amount of water applied to crops with GPS-guided irrigation systems. With the use of this approach, variable-rate irrigation—which applies water in varied amounts to different parts of a field according to their individual needs—is made possible. This keeps water resources from being overused by ensuring that crops receive the right amount of water. Furthermore, soil moisture levels can be measured in real-time using sensors, and irrigation schedules can be modified accordingly to avoid under-irrigation and guarantee stress-free crops. Additionally, it offers variable-rate pesticide and fertiliser application throughout a field, tailored to the unique requirements of each region. This is achieved by using GPS technology to map the field and identify areas with varying soil conditions, crop health, and pest or weed pressure. Fertilizers and pesticides can be applied more precisely to areas where they are most needed, reducing the overall amount of pesticide used and minimizing environmental impact.

The fundamental set back of these techniques are high set up and initial cost, requirement of technical skills, and market constrains.

Sustainable Diversification

It is a set of techniques practiced to maintain soil health by diversified crops. It involves the following measures.

Crop rotation: It is an essential farming technique that entails growing various crops on the same plot of land in a predetermined order. By using this technique, crop yields are increased overall, pest and disease pressure is decreased, and soil health is maintained. Crop rotation is widely used, especially in developed nations and many other locations. This is because supporting technology are available and people are becoming more aware of its advantages. Three-crop rotation, two-crop rotation, and lengthier rotations are examples of common crop rotation systems. Three-crop rotations comprise three crops, such as corn, soybeans, and wheat, while two-crop rotations alternate two crops, such as corn and soybeans or wheat and barley. Longer crop rotations with four or more crops may be used in some areas to increase soil health and diversification. Crop rotation improves soil structure, and reduce soil erosion, reduce pest and disease pressure by disrupt life cycle of pests, balance nutrient uptake and prevent nutrient depletion and suppress weeds more effectively.

Intercropping: It's a farming technique in which multiple crops are produced concurrently on one plot of land. A few advantages of this approach include higher yields, better soil health, and less stress from pests and diseases. However precise data regarding the use and advantages of intercropping might differ significantly based on location, crop varieties, and agricultural techniques. Intercropping is becoming more and more popular, especially in areas where there is a need for more food production or where there are limited land resources. As a result, intercropping is more common in poorer nations where it is a customary farming method that can enhance food security. Legumes with cereals, cover crops with crops, and tree-crop systems are examples of common intercropping systems. Legumes, like soybeans or lentils, are often planted alongside cereals, like corn or wheat, in an intercropping system. To enhance soil health and lower erosion, cash crops can be interplanted with cover crops like winter rye or vetch. Tree-crop systems consist of trees that can be planted alongside agricultural crops to create shade, increase soil fertility, and increase revenue diversification. Comparing intercropping versus monoculture, overall yields might frequently be higher. enhanced biodiversity, decreased pest and disease pressure, improved soil organic matter, improved soil structure, decreased soil erosion, and optimal use of resources like water and nutrients.

Agroforestry: It's a kind of land-use management where crops are grown on the same plot of land as trees and plants. Increased biodiversity, better soil health, and increased climate change resilience are just a few advantages of this method. The benefits and adoption of agroforestry might differ significantly based on the geography, crop varieties, and agricultural techniques. Agroforestry is becoming more and more popular, especially in areas where there are limited land resources or where greater sustainability is required. In underdeveloped nations, where it is a customary agricultural method and can enhance food security and livelihoods, it is frequently more common. Home gardens, tree-crop systems, and forest farms are examples of common agroforestry systems. Small-scale agroforestry systems called "home gardens," which combine fruit trees, vegetables, and other crops, are common in rural areas. Tree-crop systems consist of trees that can be grown alongside agricultural products, such cocoa or coffee, to increase soil fertility, create shade, and increase revenue diversification. Furthermore, forest farms are larger-scale agroforestry projects that grow trees on agricultural land in an effort to improve the landscape's resilience and sustainability. Agroforestry offers protection from extreme weather, reduces water loss, and provides shade, all of which can assist to lessen the effects of climate change. It can lessen farmers' dependency on a particular crop by giving them access to several revenue streams. By supplying food, fuel, and cash, agroforestry can help rural populations live better lives. However, the initial cost of implemention, planning and management, and market constrains can be a barrier for some farmers.

Conservation techniques

To reduce the economic burden of agriculture especially in developing high populated countries conservation techniques are more promising. It includes techniques like no-till farming, cover-croping, water conservation and efficient irrigation. Following are the type of conservation techniques.

No-till farming: This is not the same as traditional tillage methods, which include harrowing or ploughing the soil in order to get it ready for planting. This type of farming lowers soil erosion, increases the amount of organic matter in the soil, and maintains the soil's structure. Compaction of the soil caused by tillage can hinder root penetration and decrease water infiltration. Compaction of the soil is less common with no-till farming. No-till farming can help retain soil moisture by minimising soil disturbance. It can also increase biodiversity by providing home for beneficial insects and other species and reduce greenhouse gas emissions by lowering soil carbon loss. This is especially true in dry locations. Because they might not be disturbed by tillage, weeds can be a problem in no-till systems. Nevertheless, tillage need not be used in order to control weeds when using efficient weed management techniques. This method also necessitates a large initial investment and cautious handling of agricultural wastes. In many parts of the world, no-till farming is becoming more and more popular; adoption rates have risen recently.

Cover cropping: Plants are cultivated in between the primary income crops as a cover crop to improve soil health, lower erosion, and increase biodiversity. Grass and legumes, or a mix of the two, can be used as cover crops. In recent years, cover crops have become more and more popular, especially with farmers looking to reduce their dependency on synthetic inputs and enhance soil health. In industrialised nations, when access to auxiliary technology and a heightened understanding of its advantages are prevalent, it is more widespread. Brassicas like mustards and radishes can help to break up compacted soil and suppress weeds. Grasses like ryegrass, oats, and sorghum can help to reduce soil erosion and improve soil structure. Legumes like soybeans, peas, and clover can improve soil nitrogen content and draw beneficial insects. Increased nutrient cycling, less pest and disease pressure, better soil organic matter, improved soil structure, decreased soil erosion, increased biodiversity, and improved water quality are all benefits of cover crops. However, there are commercial limitations and a high initial cost associated with the procedure.

Water conservation: implementation of effective watering methods, such drip irrigation, and making use of rainwater collection strategies. Conserving water is vital, particularly in areas that are experiencing drought or water scarcity. Effective water management is essential for irrigation in order to maintain sustainable farming methods. The following are a few efficient irrigation methods for conserving water: water harvesting, drip irrigation, sprinkler irrigation, soil moisture sensors (which use real-time monitoring to detect soil moisture levels and enable farmers to irrigate only when needed), and efficient irrigation scheduling.

Mulching: It can be characterised as an evaporation reduction technique. Mulching is a gardening and agricultural practice that entails adding organic or inorganic materials to the soil surrounding plants. Depending on the type of mulch and the plants being grown, there are several appropriate depths for mulch. Usually, 2-4 inches of layer is enough. Moisture retention, weed control, soil temperature regulation, erosion prevention, nutrient enhancement, and aesthetics are just a few advantages of this method. Mulch is classified into two groups. The first type of mulch is organic, which is made of things like compost, wood chips, straw, grass clippings, and leaves. As organic mulch breaks down over time, it enriches the soil with nutrients. Second, materials like rocks, pebbles, landscape fabric, and plastic sheeting are examples of inorganic mulch. Since it doesn't break down, it's usually applied for long-term weed management. In addition, enhancing drainage also keeps the area from being too wet, which avoids nutrient loss and root rot. Additionally, selecting crop kinds resistant to drought helps cut down on water usage. Furthermore, certain crops are inherently more water-efficient than others, so choose carefully.

Soil Health

Organic matter: It is essential to the health of soils. It is essential for strengthening nutrient availability, retaining more water in the soil, boosting microbial activity, and improving soil structure. The addition of organic matter to soil enhances its porosity and structure, leading to better root development, drainage, and water infiltration. It lessens the requirement for frequent irrigation by increasing the soil's ability to hold water. Beneficial soil microorganisms, which are essential to the cycling of nutrients, the prevention of illness, and the health of the soil, eat it. Additionally, it increases soil fertility and binds soil particles together, lowering the danger of erosion and improving the quality and production of crops.

Nutrient management: maximising the application of fertiliser to satisfy crop requirements while reducing nutrient outflow. An essential component of sustainable agriculture is nutrient management, which focusses on using fertilisers effectively and efficiently to meet crop nutrient requirements while reducing environmental impact. The nutrients that plants most need are nitrogen, potassium, phosphorus, calcium, sulphur, and manganese. The amount of nutrients in the soil is influenced by the type of soil, climate, crop rotation, and fertiliser. Testing the soil on a regular basis aid in assessing its nutritional content and identifying surpluses or deficits. Neutralising nutrient levels in the crop can be achieved by applying fertilisers in the right amounts to suit its unique requirements. Nutrient utilisation can be maximised by applying fertilisers at the proper rate and location using precision agricultural techniques. Water quality can be preserved and nutrients can be preserved by putting into practice strategies to lessen nutrient discharge, leaching, and volatilisation. The availability and retention of nutrients in the soil can be enhanced by adding organic matter. Crop rotation can lower the chance of nutrient deficits and preserve soil fertility.

Optimum use of fertilizer: It includes Right fertilizer, timing, rate, placement of fertilizer, Avoidance of nutrient losses by to prevent nutrient losses through runoff, leaching, and volatilization.

Reduced environmental impacts: Reducing greenhouse gas emissions and safeguarding water quality can both be achieved by minimising nutrient losses. evaluating the soil frequently to identify nutrient deficits and modifying fertiliser applications as necessary. It offers insightful information on the soil's nutrient state, enabling farmers to apply fertiliser wisely, avoiding overfertilization and guaranteeing crops get the nutrients they need. Furthermore, frequent soil testing promotes environmental sustainability by lowering pollution and nutrient discharge, which both have an adverse effect on the ecosystem.

Pest and Disease Management

For agricultural practices to be sustainable, insect control is crucial. Insects like aphids, caterpillars, beetles, and weevils, as well as microorganisms like fungi, bacteria, and viruses, are common pests for crops. One of the most successful methods of controlling pests is integrated pest management, or IPM. It entails eliminating dangerous organisms that could harm crops and lower production. Its goal is to keep pests under control with the least amount of chemical pesticides. In order to manage pests, biocontrol uses natural enemies like nematodes, microbes, and beneficial insects. Effective biocontrol also involves growing crop varieties that are naturally resistant to pests and illnesses. Additionally, cultural control is a method that makes use of agricultural techniques like crop rotation, cover crops, and sanitation that can lessen the burden of pests and diseases. Furthermore, physical control—using physical barriers or techniques like row coverings, netting, or traps to keep pests away from crops—is both fairly common and reasonably priced. To reduce the negative effects on the environment, it's critical to apply pesticides sparingly and in accordance with label recommendations. Integrated pest management strategies reduce expenses on chemical pesticides. By implementing effective pest and disease management strategies, farmers can protect their crops, maintain soil health, and ensure sustainable agricultural production.

Policy and Governance

One of the eight missions delineated under the National Action Plan on Climate Change (NAPCC), which was suggested in 2008, was the national mission for sustainable agriculture (NMSA) in India. It is the main initiative of the central government and has been a programmatic intervention from 2014 to 2015 with the goal of making agriculture sustainable on all fronts. Land reform guarantees smallholder farmers access to resources and a fair allocation of land. To guarantee the sustainable use of agricultural land, effective governance and policy are crucial. In addition to addressing the issues of food security and environmental protection, these policies and governance frameworks can support sustainable practices and the preservation of agricultural resources. It covers environmental control, land tenure security, biodiversity conservation in agricultural land, water resource management, land use planning, and agricultural subsidies. The Indian government's programs for sustainable farming practices are listed below.

- Development of rainfed areas: To increase farmer incomes and lessen the effects of floods and drought, the program encourages integrated agricultural systems in rainfed areas of the nation that incorporate diversified cropping techniques in addition to horticulture, cattle, fisheries, etc.
- Submission on agroforestry: This mission, which was initiated in 2016 in response to the National Agroforestry Policy's adoption, attempts to promote tree planting in a way that is connected with crop production (MoAFW 2016).

- Soil health management: This important program aims to "create and link soil fertility maps with macro-micro nutrient management, judicious application of fertilisers and organic farming practices, as well as location as well as crop specific sustainable soil health management."
- In 2015, the Paramparagat Krishi Vikas Yojna (PKVY) was introduced to develop organic agriculture in India.
- The Pradhan Mantra Krishi Shichayee Yojana (PMKSY) luncheon in 2015 is centred on increasing the nation's irrigation infrastructure, especially in rainfed areas, and enhancing agricultural water usage efficiency.
- The Pradhan Mantra Fasal Bima Yojana (PMFBY) was introduced in 2016 and offers farmers crop insurance coverage from the time of sowing until the harvest to compensate for losses due to unavoidable natural disasters.

Adoption of sustainable agriculture practices is influenced by a variety of factors. For example, older farmers, illiteracy, gender, and socioeconomic status all provide challenges to the adoption of innovative farming techniques. It has been shown that younger, more educated women farmers with higher incomes are better at implementing new technologies than their peers. Larger farmers also experiment with new technologies in certain areas of their farms.

Conclusion

Sustainable farming practices shield the land, water, biodiversity, and downstream resources from harm, as well as the people who live or work nearby or on the farm. High costs, the need for expertise and space, as well as market limitations, are some of the main obstacles to the adoption of sustainable agriculture practices in underdeveloped nations. Adopting new technologies also requires farmer understanding and government legislation.

References

Almond, R.E.A., Grooten, M., and Peterson, T. (2020). Living Planet Report 2020-Bending the curve of biodiversity loss. World Wildlife Fund.

Aryal, J.P., Khatri-Chhetri, A., Sapkota, T.B., Rahut, D.B., and Erenstein, O. (2020). Adoption and economic impacts of laser land leveling in the irrigated rice-wheat system in Haryana, India using endogenous switching regression. Natural Resources Forum, 44(3), 255 273. DOI:10.1111/1477-8947.12197.

Coulibaly T.P., Du J. and Diakité D. (2021). Sustainable agricultural practices adoption. Agriculture (Po nohospodárstvo). 67(4), 166 – 176.

Harwood, R.R. (2020). A history of sustainable agriculture. In Edwards, C.A., Lal, R., Madden, P., Miller, R.H. and House, G. (Eds.) Sustainable Agricultural Systems, pp. 3–19. Boca Raton: CRC Press. DOI:10.1201/9781003070474.

Li, F., Zhang, K., Ren, J., Yin, C., Zhang, Y., and Nie, J. (2021). Driving mechanism for farmers to adopt improved agricultural systems in China: The case of rice-green manure crops rotation system. Agricultural Systems, 192, 103202. doi:10.1016/j.agsy.2021.103202.

Negassa, M.D., Mallie, D.T., and Gemeda, D.O. (2020). Forest cover change detection using Geographic Information Systems and remote sensing techniques: A spatio-temporal study on Komto Protected forest priority area, East Wollega Zone, Ethiopia. Environmental Systems Research, 9(1), 1. DOI:10.1186/s40068-020-0163-z.

NITI Aayog, (2019). "Achievements in the year 2018-19". https://niti.gov.in/verticals/land-and-water-resources/achievements-of-the-year-2018-19. OECD/ICRIER, (2018). Agricultural Policies in India. OECD Food and Agricultural Reviews. OECD Publishing: Paris.

Nyanga, P.H., Umar, B.B., Chibamba, D., Mubanga, K.H., Kunda-Wamuwi, C.F., and Mushili, B.M. (2020). Chapter 6 Reinforcing ecosystem services through conservation agriculture in sustainable food systems. In Rusinamhodzi, L. (Ed.) The Role of Ecosystem Services in Sustainable Food Systems, Academic Press, pp. 119–133. DOI:10.1016/B978-0-12-816436-5.00006-8.

Siebrecht, N. (2020). Sustainable agriculture and its implementation gap Overcoming obstacles to implementation. Sustainability, 12(9), 3853. DOI:10.3390/su12093853.

Yanakittkul, P. and Aungvaravong, C. (2020). A model of farmers intentions towards organic farming: A case study on rice farming in Thailand. Heliyon, 6(1), e03039.

CHAPTER 7

TOWARDS A SUSTAINABLE FUTURE: EFFECTIVE STRATEGIES FOR ADDRESSING CLIMATE CHANGE

Dr. Roshan Lal Sahu

Department of Botany, Government Naveen College Kamleshwarpur, Mainpat, Surguja C.G. (roshan.lal1895@gmail.com)

Abstract

Climate change, driven by greenhouse gas emissions from human activities such as industrial processes, deforestation, and fossil fuel consumption, poses significant threats to ecosystems and human societies. The chapter outlines the current impacts, including biodiversity loss, ocean acidification, and the increasing frequency of extreme weather events, which threaten food security, water resources, and public health. The importance of adaptation strategies is highlighted, focusing on resilient infrastructure, ecosystem-based approaches, and public health adaptation. These strategies aim to minimize vulnerabilities and enhance resilience in the face of climate change. The role of communities and governance in climate adaptation planning is also discussed, emphasizing the need for participatory approaches and international cooperation. The paper underscores the significance of global initiatives such as the Paris Agreement and the Green Climate Fund in supporting adaptation efforts, particularly in vulnerable regions. By integrating sustainable practices and fostering community engagement, the paper advocates for a coordinated global response to mitigate climate impacts and build a resilient, sustainable future.

Keywords: Climate Change, Green House Gas, Deforestation, Effective Strategies For Climate Change, Resilient Infrastructure, Ecosystem.

Introduction

Climate change refers to long-term alterations in temperature, precipitation patterns, and other climatic variables that result from human activities, primarily the emission of greenhouse gases (GHGs) such as carbon dioxide (CO_2), methane (CH_4), and nitrous oxide (N_2O) (Hui et al., 2022). These gases trap heat in the Earth's atmosphere, leading to a gradual warming of the planet—an effect commonly referred to as the greenhouse effect. Key drivers of climate change include industrial processes, fossil fuel combustion for energy production, deforestation, agriculture, and land-use changes (Dale et al., 2011). These activities release vast amounts of GHGs into the atmosphere, accelerating the natural greenhouse effect and causing global temperatures to rise. The consequences of climate change are far-reaching, affecting ecosystems, weather patterns, sea levels, and human societies worldwide (Upadhyay, 2020). Mitigating climate change requires concerted efforts to reduce GHG emissions, transition to renewable energy sources, conserve forests, and adopt sustainable practices in agriculture and urban planning (Elum et al., 2017). Climate change continues to exert profound and escalating impacts on both natural ecosystems and human societies globally. In recent years, the evidence of these impacts has become increasingly clear and alarming. Ecosystems are experiencing shifts in distribution and composition, affecting biodiversity and ecosystem services crucial for human well-being. For instance, coral reefs are suffering from bleaching events due to rising sea temperatures, while species are migrating to higher altitudes or latitudes in search of suitable habitats (Munday et al., 2008). These changes reverberate through human societies, disrupting livelihoods dependent on agriculture, fisheries, and forestry. Extreme weather events such as hurricanes, floods, and droughts are becoming more frequent and intense, exacerbating food insecurity and displacing vulnerable communities (Bell & Masys, 2020). In urban areas, heatwaves pose health risks, especially to the elderly and marginalized populations. Looking ahead, projections indicate that without substantial mitigation efforts, these impacts will intensify (Chen et al., 2023). Coastal cities face increased risks from sea-level rise, while agricultural productivity is expected to decline in many regions (Nicholls et al., 2015). Social inequalities are likely to widen as poorer communities bear the brunt of climate impacts. Addressing climate change requires urgent and coordinated global action to mitigate greenhouse gas emissions and adapt to inevitable changes. The stakes are high, and the time for decisive action is now to safeguard both ecosystems and societies for future generations.

Current Impacts

Ecosystems

Biodiversity loss, ocean acidification, glacier retreat are major impact on ecosystems (Fell et al., 2017). Species are experiencing habitat loss, shifting ranges, and altered reproductive cycles due to changing temperatures and precipitation patterns (Upadhyay, 2020). Increased CO_2 absorption is causing oceans to become more acidic, threatening marine life like corals and shellfish (Abbasi and Abbasi,

2011). On the other hand, glaciers are melting at unprecedented rates, affecting water availability for communities dependent on glacial meltwater (Sharma and Sharma, 2008).

Societies

Food Security, Water Resources, Health Impacts are major impact on societies. Changing climate patterns are affecting crop yields and livestock productivity, leading to food shortages and price volatility (Fonta et al., 2011). Changes in precipitation patterns and melting glaciers are altering water availability, impacting agriculture, drinking water supplies, and sanitation (Grover, 2015). Moreover, Increased frequency of extreme weather events (like heatwaves, floods, and storms) and changing disease patterns (e.g., vector-borne diseases) pose risks to human health (Gulzar et al., 2021).

Mitigation and Adaptation Efforts

Governments, organizations, and communities are taking steps to mitigate and adapt to these impacts through various strategies. Reducing greenhouse gas emissions through policies, regulations, and technological innovations aimed at transitioning to renewable energy and sustainable practices (Jordaan et al., 2017). Developing strategies to cope with and minimize the impacts of climate change, such as building resilient infrastructure, implementing water management strategies, and enhancing disaster preparedness (Jabareen, 2013).

Adaptation Strategies

Adaptation strategies are essential responses to the challenges posed by climate change, focusing on minimizing vulnerability and enhancing resilience in ecosystems, economies, and societies. As the impacts of global warming become increasingly evident, adaptation becomes a critical complement to mitigation efforts aimed at reducing greenhouse gas emissions. One key area of adaptation is in agriculture, where farmers are deploying practices like crop diversification, improved water management, and the use of drought-resistant seeds to mitigate the effects of changing precipitation patterns and temperature extremes (Altieri et al., 2015). In coastal regions, adaptation strategies include the construction of sea walls, mangrove restoration, and sustainable land-use planning to protect against sea-level rise and storm surges (Haque et al., 2019). Urban areas are also implementing adaptation measures such as green infrastructure (like parks and green roofs) to combat heat islands and enhance water management capabilities (Wong et al., 2021). Additionally, cities are developing early warning systems for extreme weather events to improve disaster preparedness and response (Tang et al., 2012). Community involvement is crucial in successful adaptation efforts, ensuring that local knowledge and needs are integrated into adaptation planning and implementation. Policies that support adaptation, such as incentives for climate-resilient infrastructure and building codes that account for future climate risks, are also vital (Vallejo and Mullan, 2017). Ultimately, effective adaptation requires collaboration across sectors and scales, from local communities to national governments and international organizations. By investing in adaptation strategies now, societies can reduce the impacts of climate change, protect vulnerable populations, and foster sustainable development pathways in a changing world.

A. Infrastructure Resilience

Resilient infrastructure refers to the ability of infrastructure systems to anticipate, absorb, adapt to, and rapidly recover from disruptions, including those caused by climate change (Kumar et al., 2021). This type of infrastructure is crucial for ensuring sustainable development and the well-being of communities and economies in the face of increasing environmental challenges. Resilient infrastructure reduces vulnerability to climate-related hazards such as floods, storms, and heatwaves, thereby minimizing economic losses and preserving lives (Le et al., 2020). On the other hand, adaptation helps communities adapt to changing climate conditions by integrating climate resilience into infrastructure design, construction, and maintenance (Le et al., 2020). Morover, Resilient infrastructure ensures that investments in infrastructure are sustainable and can withstand the impacts of climate change over their lifecycle. Green infrastructure projects that incorporate natural systems, such as green roofs, permeable pavements, and urban forests, to manage stormwater, reduce urban heat island effects, and enhance biodiversity (Pitman et al., 2015). For example, Chicago's green roofs ordinance promotes the installation of green roofs to mitigate urban heat and manage stormwater (Carter and Fowler, 2008). Infrastructure designed to manage flood risks, such as flood barriers, levees, and drainage systems. The Thames Barrier in London is a notable example of infrastructure designed to protect the city from storm surges and rising sea levels (Hall et al., 2019). Morover, infrastructure projects that promote renewable energy sources like solar and wind power, reducing dependence on fossil fuels and mitigating greenhouse gas emissions (Mac Kinnon etal., 2018). The Masdar City project in Abu Dhabi aims to be carbon-neutral and is a model of sustainable urban development (Reiche, 2010).Top of FormBottom of Form

B. Ecosystem-based Adaptation

Ecosystem-based adaptation (EbA) measures leverage natural ecosystems and biodiversity to help communities adapt to climate change impacts (Littlefield et al., 2019). These approaches offer multiple benefits by enhancing ecosystem resilience and supporting sustainable livelihoods. Restoring mangrove forests along coastlines can provide natural barriers against storm surges and sea-level rise (Haque et al., 2019). Mangroves stabilize shorelines, reduce erosion, and provide habitat for marine life, benefiting both coastal communities and biodiversity (Haque et al., 2019). Besides, integrating trees and shrubs into agricultural landscapes can improve soil fertility, enhance water retention, and provide shade for crops (Grover, 2015). Agroforestry systems help farmers adapt to changing climate conditions by diversifying their income sources and reducing vulnerability to droughts and floods (Awazi, 2022). Moreover, sustainable forest management practices, including reducing deforestation, promoting reforestation, and implementing fire prevention measures, help maintain forest resilience (Braat, 2012). Healthy forests contribute to carbon sequestration, regulate water cycles, and provide habitat for wildlife (Braat, 2012). On the other hand, green infrastructure in urban areas, coral reef conservation, ecosystem-based water management, agricultural adaptation, disaster risk reduction are important facotr for ecosystem based adaptation (McVittie et al., 2018).

C. Public Health Adaptation

Climate change poses significant and varied health impacts, affecting populations worldwide through direct and indirect mechanisms. Increasing temperatures and more frequent heatwaves can lead to heat stress, heat exhaustion, and heatstroke, particularly affecting vulnerable populations such as the elderly, children, and outdoor workers (Kravchenko et al., 2013). Climate change intensifies extreme weather events such as hurricanes, floods, and wildfires, leading to injuries, displacement, and mental health issues among affected populations. Climate change exacerbates air pollution by increasing ground-level ozone and particulate matter concentrations, which can worsen respiratory and cardiovascular diseases and increase mortality rates, especially in urban areas (Sheffield et al., 2011). Changes in temperature and precipitation patterns can alter the distribution and seasonality of vector-borne diseases like malaria, dengue fever, and Lyme disease, expanding their geographic range and increasing transmission risks. Changes in precipitation patterns and rising temperatures can impact water quality and availability, increasing the risk of waterborne diseases such as cholera and diarrheal diseases (Levy et al 2016). Climate change affects agricultural productivity and food availability, leading to malnutrition, food insecurity, and related health problems, especially in low-income and vulnerable communities (Elum et al., 2017). Climate change impacts can exacerbate social inequalities, displacement, and migration, which in turn can lead to social disruption, violence, and mental health challenges (McMichael et al., 2010). Effective policies and interventions to protect public health from the impacts of climate change involve a combination of adaptive measures, community engagement, and robust health systems. Public health adaptation measures to mitigate climate change include enhancing heatwave preparedness, improving water management to combat droughts and floods, promoting vector control to prevent diseases like malaria and dengue fever, ensuring food security through resilient agricultural practices, and bolstering healthcare infrastructure to handle increased health risks (Levy et al 2016). Additionally, raising public awareness about climate-related health impacts and fostering community resilience are crucial. These measures aim to protect vulnerable populations, mitigate health inequalities exacerbated by climate change, and build adaptive capacity to ensure sustainable public health outcomes in the face of changing environmental conditions ((Levy et al 2016)).

Community Engagement and Governance

Communities play a crucial role in climate change adaptation planning and implementation due to their intimate knowledge of local conditions, vulnerabilities, and resources. Here are several key roles communities can play in climate change adaptation:

1. Identifying Local Risks and Priorities

Communities are often the first to experience and understand climate impacts specific to their region, such as floods, droughts, or heatwaves. They can identify vulnerabilities in infrastructure, livelihoods, and ecosystems. By engaging with communities, planners can gain insights into the priorities and concerns of local residents regarding climate impacts (Corburn, 2003).

2. Participatory Planning and Decision-Making

Involving communities in adaptation planning ensures that solutions are locally appropriate and supported. Participatory approaches foster ownership and commitment to adaptation measures. Community members can contribute traditional knowledge, cultural practices, and innovative ideas that complement scientific approaches to adaptation (Corburn, 2013).

3. Implementing and Monitoring Adaptation Actions

Communities are instrumental in implementing adaptation actions on the ground, such as constructing flood defenses, implementing water management strategies, or promoting sustainable agricultural practices (Nkoana et al., 2018). Local knowledge supports the monitoring and evaluation of adaptation measures, ensuring they remain effective and adaptable to changing conditions.

4. Education and Awareness

Community engagement promotes education and awareness about climate change impacts and adaptation strategies. This empowers individuals to take proactive measures to protect themselves and their communities (Narksompong and Limjirakan, 2015). Education initiatives can include training programs on sustainable practices, disaster preparedness, and climate-smart technologies.

National Climate Change Adaptation Plans (NAPs)

NAPs outline a country's strategy and actions for adapting to climate change impacts (Laukkonen et al., 2009). They typically involve assessing vulnerabilities, setting adaptation priorities, and outlining implementation strategies. NAPs provide a coordinated approach to adaptation across sectors and levels of government, ensuring that adaptation measures are integrated into national development plans (Mizuno and okano, 2024). Sector-specific policies (e.g., water management, agriculture, health) integrate climate considerations into planning and regulation, ensuring resilience against climate impacts such as droughts, floods, and water scarcity (Mizuno and okano, 2024). Local governments develop climate action plans tailored to local vulnerabilities and priorities. These plans often include measures for infrastructure upgrades, land-use planning, and community engagement to enhance resilience at the local level (King et al., 2016). Besides, international agreements establish frameworks for global cooperation on climate change mitigation and adaptation. They promote knowledge sharing, capacity building, and financial support for developing countries' adaptation efforts (Khan et al., 2021). On the other hand, collaboration between governments, businesses, and civil society organizations fosters innovation and resource mobilization for climate adaptation (Khan et al., 2021). PPPs (Public-Private Partnerships) can leverage private sector expertise and investments to scale up adaptation measures (Mizuno and okano, 2024).

Case Study Example: National Adaptation Plan (NAP) Process in Uganda

Uganda's NAP process involves coordination across multiple government ministries, departments, and agencies under the guidance of the Climate Change Department within the Ministry of Water and Environment. The NAP outlines priority sectors (e.g., agriculture, water resources, health) and integrates adaptation actions into national development planning processes. Uganda's NAP focuses on enhancing resilience through measures such as climate-smart agriculture, improved water management, and community-based adaptation initiatives (Nyasimi et al., 2016). The process includes monitoring of climate impacts, adaptation actions, and capacity-building efforts, with periodic reviews to adjust strategies based on emerging climate risks and evolving needs (Nyasimi et al., 2016).Top of FormBottom of Form

International Cooperation and Policy Development

Global initiatives and agreements on climate adaptation play a crucial role in coordinating efforts across countries to address the impacts of climate change and build resilience in vulnerable communities. Here are some key initiatives and agreements:

1. **Paris Agreement:** Adopted in 2015 under the United Nations Framework Convention on Climate Change (UNFCCC), the Paris Agreement aims to limit global warming to well below 2 degrees Celsius above pre-industrial levels, with efforts to limit the temperature increase to 1.5 degrees Celsius (Delbeke et al., 2019). While primarily focused on mitigation, the agreement also emphasizes adaptation as a key component (Delbeke et al., 2019). Parties are encouraged to enhance adaptive capacity, strengthen resilience, and reduce vulnerability to climate change.
2. **Sendai Framework for Disaster Risk Reduction:** Adopted in 2015 at the Third UN World Conference on Disaster Risk Reduction, the Sendai Framework emphasizes the importance of integrating disaster risk reduction

and resilience-building into national policies and strategies (Aitsi-Selmi et al., 2019). It includes specific targets for reducing disaster mortality, economic losses, and damage to critical infrastructure by promoting investment in resilience and adaptive practices.

3. **UNFCCC Adaptation Framework:** The UNFCCC has established various mechanisms and frameworks to support adaptation actions, including the Adaptation Committee and the National Adaptation Plans (NAPs) process (Morgan and Waskow, 2014). The Adaptation Committee provides technical support and guidance to countries on adaptation planning and implementation, while NAPs help countries identify their adaptation needs and prioritize actions (Morgan and Waskow, 2014).
4. **Green Climate Fund (GCF):** The GCF is a financial mechanism under the UNFCCC designed to support developing countries in their efforts to mitigate and adapt to climate change. It provides funding for adaptation projects and programs that enhance resilience and reduce vulnerability, particularly in the most vulnerable and least developed countries (Afful-Koomson, 2015).
5. **The Global Commission on Adaptation (GCA):** Launched in 2018, the GCA is an initiative co-chaired by Ban Ki-moon, Bill Gates, and Kristalina Georgieva, aiming to accelerate adaptation actions and elevate the political visibility of adaptation. The GCA promotes the integration of adaptation into economic development planning and investment decisions at all levels (Szilagyi, 2020).
6. **Regional and Bilateral Agreements:** Many regions and countries have also developed their own adaptation initiatives and agreements. For example, the African Union has adopted the Africa Adaptation Initiative to enhance adaptation efforts across the continent (Dzebo, 2019), while the European Union's adaptation strategy aims to foster climate-resilient societies and economies within Europe (Oberthür and Dupont, 2021). These global initiatives and agreements provide frameworks, financial support, and technical assistance to help countries enhance their adaptive capacity and resilience. On the other hand, international organizations and partnerships play a pivotal role in addressing climate change by fostering collaboration, providing technical expertise, mobilizing funding, and promoting policy coherence among countries (Dzebo, 2019).

Conclusion

In conclusion, addressing climate change demands urgent global action across all sectors to mitigate its impacts and build resilience. The evidence is clear, ecosystems are under strain, societies face heightened risks to food security and health, and vulnerable communities are disproportionately affected. Effective strategies include reducing greenhouse gas emissions, transitioning to renewable energy, conserving ecosystems, and integrating climate resilience into infrastructure and urban planning. Adaptation measures, such as enhancing agricultural practices, protecting natural barriers like mangroves, and improving public health preparedness, are essential. International cooperation through agreements like the Paris Agreement and initiatives such as the Green Climate Fund are crucial for supporting adaptation efforts worldwide. By investing in sustainable practices and fostering community engagement, we can navigate towards a more resilient and sustainable future for all.

References

Abbasi, T., & Abbasi, S. A. (2011). Ocean acidification: The newest threat to the global environment. *Critical Reviews in Environmental Science and Technology, 41*(18), 1601-1663.

Afful-Koomson, T. (2015). The Green Climate Fund in Africa: what should be different?. *Climate and Development, 7*(4), 367-379.

Aitsi-Selmi, A., Egawa, S., Sasaki, H., Wannous, C., & Murray, V. (2015). The Sendai framework for disaster risk reduction: Renewing the global commitment to people's resilience, health, and well-being. *International journal of disaster risk science, 6*, 164-176.

Altieri, M. A., Nicholls, C. I., Henao, A., & Lana, M. A. (2015). Agroecology and the design of climate change-resilient farming systems. *Agronomy for sustainable development, 35*(3), 869-890.

Awazi, N. P. (2022). Agroforestry for climate change adaptation, resilience enhancement and vulnerability attenuation in smallholder farming systems in Cameroon. *Journal of Atmospheric Science Research, 5*(1), 25-33.

Bell, C., & Masys, A. J. (2020). Climate change, extreme weather events and global health security a lens into vulnerabilities. *Global health security: Recognizing vulnerabilities, creating opportunities*, 59-78.

Braatz, S. (2012). Building resilience for adaptation to climate change through sustainable forest management. *Building resilience for adaptation to climate change in the agriculture sector, 23*, 117.

Carter, T., & Fowler, L. (2008). Establishing green roof infrastructure through environmental policy instruments. *Environmental management, 42*, 151-164.

Chen, M., Chen, L., Zhou, Y., Hu, M., Jiang, Y., Huang, D., ... & Xian, Y. (2023). Rising vulnerability of compound risk inequality to ageing and extreme heatwave exposure in global cities. *npj Urban Sustainability, 3*(1), 38.

Corburn, J. (2003). Bringing local knowledge into environmental decision making: Improving urban planning for communities at risk. *Journal of planning education and research, 22*(4), 420-433.

Dale, V. H., Efroymson, R. A., & Kline, K. L. (2011). The land use–climate change–energy nexus. *Landscape ecology, 26*, 755-773.

Delbeke, J., Runge-Metzger, A., Slingenberg, Y., & Werksman, J. (2019). The paris agreement. In *Towards a climate-neutral Europe* (pp. 24-45). Routledge.

Dzebo, A. (2019). Effective governance of transnational adaptation initiatives. *International Environmental Agreements: Politics, Law and Economics, 19,* 447-466.

Elum, Z. A., Modise, D. M., & Nhamo, G. (2017). Climate change mitigation: the potential of agriculture as a renewable energy source in Nigeria. *Environmental Science and Pollution Research, 24,* 3260-3273.

Fell, S. C., Carrivick, J. L., & Brown, L. E. (2017). The multitrophic effects of climate change and glacier retreat in mountain rivers. *BioScience, 67*(10), 897-911.

Fonta, W., Edame, G., Anam, B. E., & Duru, E. J. C. (2011). Climate change, food security and agricultural productivity in Africa: Issues and policy directions.

Grover, V. I. (2015). Impact of climate change on the water cycle. *Managing Water Resources under Climate Uncertainty: Examples from Asia, Europe, Latin America, and Australia,* 3-30.

Gulzar, A., Islam, T., Gulzar, R., & Hassan, T. (2021). Climate change and impacts of extreme events on human health: An overview. *Indonesian Journal of Social and Environmental Issues (IJSEI), 2*(1), 68-77.

Hall, J. W., Harvey, H., & Manning, L. J. (2019). Adaptation thresholds and pathways for tidal flood risk management in London. *Climate Risk Management, 24,* 42-58.

Haque, A., Rahman, M. H., & Rahman, D. (2019). An evaluation of sea level rise vulnerability and resilience strategy to climate change in the coastline of Bangladesh. *International Journal of Environmental Sciences and Natural Resources, 18*(1), 555983.

Hui, D., Deng, Q., Tian, H., & Luo, Y. (2022). Global climate change and greenhouse gases emissions in terrestrial ecosystems. In *Handbook of climate change mitigation and adaptation* (pp. 23-76). Cham: Springer International Publishing.

Jabareen, Y. (2013). Planning the resilient city: Concepts and strategies for coping with climate change and environmental risk. *Cities, 31,* 220-229.

Jordaan, S. M., Romo-Rabago, E., McLeary, R., Reidy, L., Nazari, J., & Herremans, I. M. (2017). The role of energy technology innovation in reducing greenhouse gas emissions: A case study of Canada. *Renewable and Sustainable Energy Reviews, 78,* 1397-1409.

Khan, M., Mfitumukiza, D., & Huq, S. (2021). Capacity building for implementation of nationally determined contributions under the Paris Agreement. In *Making Climate Action More Effective* (pp. 95-106). Routledge.

King, D., Gurtner, Y., Firdaus, A., Harwood, S., & Cottrell, A. (2016). Land use planning for disaster risk reduction and climate change adaptation: Operationalizing policy and legislation at local levels. *International journal of disaster resilience in the built environment, 7*(2), 158-172.

Kravchenko, J., Abernethy, A. P., Fawzy, M., & Lyerly, H. K. (2013). Minimization of heatwave morbidity and mortality. *American journal of preventive medicine, 44*(3), 274-282.

Kumar, N., Poonia, V., Gupta, B. B., & Goyal, M. K. (2021). A novel framework for risk assessment and resilience of critical infrastructure towards climate change. *Technological Forecasting and Social Change, 165*, 120532.

Laukkonen, J., Blanco, P. K., Lenhart, J., Keiner, M., Cavric, B., & Kinuthia-Njenga, C. (2009). Combining climate change adaptation and mitigation measures at the local level. *Habitat international, 33*(3), 287-292.

Le, T. M., Leow, W. J., & Seiderer, F. (2020). Building resilience in infrastructure to climate change. *Well Spent: How strong infrastructure governance can end waste in public investment*, 282-300.

Levy, K., Woster, A. P., Goldstein, R. S., & Carlton, E. J. (2016). Untangling the impacts of climate change on waterborne diseases: a systematic review of relationships between diarrheal diseases and temperature, rainfall, flooding, and drought. *Environmental science & technology, 50*(10), 4905-4922.

Littlefield, C. A. I. T. L. I. N., Nelson, E. R. I. K., Dittbrenner, B., Withey, J. O. H. N., Arkema, K., & Lawler, J. (2019). Ecosystem-based adaptation. *Biodiversity and Climate Change: Transforming the Biosphere; Lovejoy, TE, Hannah, L., Eds*, 305-317.

Mac Kinnon, M. A., Brouwer, J., & Samuelsen, S. (2018). The role of natural gas and its infrastructure in mitigating greenhouse gas emissions, improving regional air quality, and renewable resource integration. *Progress in Energy and Combustion science, 64*, 62-92.

McMichael, A., McMichael, C., Berry, H., & Bowen, K. (2010). Climate-related displacement: health risks and responses. *Climate Change and Population Displacement: Multidisciplinary Perspectives. Oxford, England: Hart Publishing Ltd.*

McVittie, A., Cole, L., Wreford, A., Sgobbi, A., & Yordi, B. (2018). Ecosystem-based solutions for disaster risk reduction: Lessons from European applications of ecosystem-based adaptation measures. *International journal of disaster risk reduction, 32*, 42-54.

Mizuno, O., & Okano, N. (2024). Reconsidering National Adaptation Plans (NAPs) as a Policy Framework under the UNFCCC. *Climate Policy*, 1-13.

Morgan, J., & Waskow, D. (2014). A new look at climate equity in the UNFCCC. *Climate Policy, 14*(1), 17-22.

Munday, P. L., Jones, G. P., Pratchett, M. S., & Williams, A. J. (2008). Climate change and the future for coral reef fishes. *Fish and Fisheries, 9*(3), 261-285.

Narksompong, J., & Limjirakan, S. (2015). Youth participation in climate change for sustainable engagement. *Review of European, Comparative & International Environmental Law, 24*(2), 171-181.

Nicholls, R. J., Reeder, T., Brown, S., & Haigh, I. D. (2015). The risks of sea-level rise for coastal cities.

Nkoana, E. M., Verbruggen, A., & Hugé, J. (2018). Climate change adaptation tools at the community level: An integrated literature review. *Sustainability, 10*(3), 796.

Nyasimi, M., Radeny, M. A., Mungai, C., & Kamini, C. (2016). Uganda's National Adaptation Programme of Action: Implementation, challenges and emerging lessons.

Oberthür, S., & Dupont, C. (2021). The European Union's international climate leadership: towards a grand climate strategy?. *Journal of European Public Policy, 28*(7), 1095-1114.

Pitman, S. D., Daniels, C. B., & Ely, M. E. (2015). Green infrastructure as life support: Urban nature and climate change. *Transactions of the Royal Society of South Australia, 139*(1), 97-112.

Reiche, D. (2010). Renewable energy policies in the Gulf countries: A case study of the carbon-neutral "Masdar City" in Abu Dhabi. *Energy policy, 38*(1), 378-382.

Sharma, B. R., & Sharma, D. (2008). Impact of climate change on water resources and glacier melt and potential adaptations for Indian agriculture. *New Dehli: International Water Management Institute.*

Sheffield, P. E., Knowlton, K., Carr, J. L., & Kinney, P. L. (2011). Modeling of regional climate change effects on ground-level ozone and childhood asthma. *American journal of preventive medicine, 41*(3), 251-257.

Szilagyi, L. (2020). Outcome validation report: CCAFS engagement in the Global Commission on Adaptation.

CHAPTER 8

ARTICLE ON QUALITY OF WORK LIFE BALANCE

Dr. Deepika Dadsena[1], Dr. Pradeep Kumar Dadsena[2]

[1]*Asst. Prof., Department of Commerce, Government Kaktiya PG College Jagdalpur, CG Email Id- deepikadarshan311@gmail.com*

[2]*Asst. Prof., Department of Mathematics, Government Engineering College Jagdalpur, CG Email Id- pradeep171744@gmail.com*

Abstract

Work Life Balance (WLB) is ability to manage resources to meet family and work demands so that individuals can how effectively participate in both domains of life. It is about effectively managing the juggling act between paid work and other activities that are important to us including spending time with family, taking part in sports and recreations volunteering or undertaking further study (Gautam & Jain, 2018). WLB is to maintain a balance between time devoted to paid work and family time. Meaning of WLB is not same for employee and employer's perspective. Employees who have a strong balance between paid work and family time are more productive and take less time off from work. They will also be happier with their jobs as they feel as though they can live their lives outside of the workplace as well as at it, without feeling drained. As per OECD data a full time worker in the OECD devotes 63% of the day on average or 15 hours to personal care and leisure and time devoted to leisure is roughly the same for men and women across the 22 OECD countries studied. The objective of this paper is to review the concept of WLB and to investigate the methods of training and development for improving the balance between work and life. Descriptive research design and secondary data has been used in the paper and the findings and conclusion will be helpful for policy makers to frame effective policy to maintain healthy and strong work life balance in the lives of employees.

Keywords: Work life balance, Mindfulness.

Introduction

WLB are the two sides of the same coin. Both work life and personal life are interconnected and interdependent. Achieving work life balance is not as simple as it sounds. It plays very important role in the modern world for men and women both. Work-life balance refers to the equilibrium between professional responsibilities and personal life activities. It encompasses managing time effectively to ensure fulfillment in both domains, without allowing one to overshadow the other. Achieving a healthy work-life balance is crucial for mental, emotional, and physical well-being, as it reduces stress, enhances productivity, and improves overall quality of life. This balance varies for individuals based on their priorities, values, and circumstances, but commonly involves setting boundaries, prioritizing tasks, and incorporating self-care practices. Striking a harmonious balance allows individuals to thrive professionally while also enjoying meaningful relationships, pursuing hobbies, and maintaining a sense of fulfillment beyond their careers. Finding equilibrium between life and work is essential for overall well-being and sustained productivity. It involves harmonizing professional commitments with personal needs and aspirations. Achieving this balance requires:

1. **Setting Priorities**: Identifying what matters most in both professional and personal spheres helps in allocating time and energy effectively.
2. **Time Management**: Efficiently planning tasks and schedules ensures that work deadlines are met while leaving adequate time for relaxation, family, and personal interests.
3. **Boundaries**: Establishing clear boundaries between work and personal life helps in preventing burnout and maintaining mental clarity.
4. **Self-Care**: Prioritizing physical exercise, adequate sleep, and mindfulness practices enhances resilience and boosts productivity.
5. **Communication**: Open communication with employers, colleagues, and family members fosters understanding and support for maintaining work-life balance.
6. **Flexibility**: Embracing flexibility in work arrangements, such as remote work or flexible hours, can facilitate better integration of work and personal commitments.

Striving for equilibrium between life and work enables individuals to lead fulfilling lives, achieve career success, and nurture meaningful relationships without compromising their overall well-being. Meaning of WLB is not same for employee and employer's perspective. From the employee view point: the dilemma of managing work obligations and personal/family responsibilities. From the employer viewpoint: the challenge of creating a supportive company culture where employees can focus on their jobs while at work. Employees who have a strong balance between paid work and family time are more productive and take less time off from work. They will also be happier with their jobs as they feel as though they can live their lives outside of the workplace as well as at it, without feeling drained. Indu Gautam and Sameeksha Jain in their conference paper 2018 found that Demographic variables like gender, age of employees, marital status educational qualification, monthly income has a significant influence on their perception towards the work life balance and work life balance has a significant impact on the family life of the employees. Further compensation factors have a very significant impact on the work life balance of employees. There is a negative correlation between work life balance and its impact on family life. i.e, if the work life of employees is imbalanced, it will create negative impact on the family life of employees as well. There is strong and positive correlation between work life balance and compensation factors i.e. compensation factor have strong influence on the work life balance of employees. The study also reveals that work environment, flexible working hours, technology, interpersonal relations with superiors, support from spouse and spending quality time with family and friends plays a very important role in their work life balance. The study concludes that the employee driven solutions for balancing work life, will help organization to design and implement work life balance policies. It will help to reduce overtime, stress and workloads on the job place and increase flexibility and family and leisure time on personal life. Welfare of employees and their satisfaction should be the topmost priority of the organizations. (Gautam & Jain, 2018).

The imbalance between these two arises when the burden, obligations and responsibilities of work and family roles become incompatible. Work life balance is the nature of women that is GOD gifted to her. Women are maintaining work life balance since long times they are considered as the manager and bosses of the family who manages every task easily. At present our finance minister Nirmala Sitaraman is managing the finance of our country very smartly, the honorable president of India Draupadi Murmu leading our country, past has full of examples where women have managed the balance between work and life. Covid-19 had given an opportunity to maintain balance between work and life, since then many people are doing online jobs but on the other side it was a big challenge also, as while engaging in household activities women have to manage to work for office. Study also reveals that there is an association between demographic variables and work life balance (Ameriya & Singh, 2021)

Finding a suitable balance between work and daily living is a challenge that all workers face. Families are particularly affected. The ability to successfully combine work, family commitments and personal life is important for the well-being of all members in a household. Governments can help to address the issue by encouraging supportive and flexible working practices, making it easier for parents to strike a better balance between work and home life.

An important aspect of work-life balance is the amount of time a person spends at work. Evidence suggests that long work hours may impair personal health, jeopardize safety and increase stress. **10% of employees in the OECD work 50 hours or more per week in paid work.** More men work very long hours in paid work;

the percentage of **male employees working very long hours in paid work across OECD countries is almost 14%, compared with about 6% for women.** Furthermore, the more people work, the less time they have to spend on other activities, such as personal care or leisure. The amount and quality of leisure time is important for people's overall well-being, and can bring additional physical and mental health benefits. As per OECD data a full time worker in the OECD devotes 63% of the day on average or 15 hours to personal care and leisure and Fewer hours in paid work for women do not necessarily result in greater leisure time, as time devoted to leisure is roughly the same for men and women across the 22 OECD countries studied.

In a recent report by the International Labor Organization (ILO) shedding light on a concerning trend, over a third of the global workforce regularly exceeds the 48-hour workweek limit. Despite some progress in promoting a more balanced lifestyle, the high incidence of burnout highlights the need for improvements across various regions. **Worst work-life balance countries** frequently face challenges such as long working hours, few days off, and increased stress from work responsibilities. In Mexico, where 27% employees work longer hours compared to the OECD average (10%). There is also a gender disparity, with 33% of men employees experiencing long working hours, and 17% of women employees are in a similar situation. In the list of the **best work-life balance countries in 2024. In** Spain, parents have the legal right to reduce their work hours to address childcare needs. This has resulted in a reduction of 18% in total work time. Although European countries dominate all listings of top countries for work-life balance, several Asian countries, such as Saudi Arabia and South Korea, also have high scores on Remote's Global Life-Work Balance Index. Several Asian companies are known for their policies that promote work-life balance. Some of the **best Asian companies for work-life balance** are Tech Mahindra Limited (NSE: TECHM), Shiseido Company, Limited (OTCMKTS: SSDOY), and Maybank (KLSE: MAYBANK).

India's Index Score is 44.32 and ranked 10th in the list of 20 Asian Countries with the Best Work-Life Balance. India has an average workweek of 31.9 hours, while the happiness index of employees stands at 3.78. The Factories Act guarantees 12 paid leave days per year for workers who have completed at least 240 working days. Moreover, workers taking 4 or more consecutive leave days receive their wages upfront, ensuring they are financially prepared for their time off.

Taiwan is a country in East Asia which stood 1st rank with *Index Score of 58.55 and Average of 29.1 Hours Worked per Week.* Taiwan's working culture emphasizes a strong work ethic, with great stress placed on work-life balance. Taiwan offers an exceptional quality of life, affordable living costs, and a world-class healthcare system accessible to all residents. This is further complemented by moderate working hours, with the standard set at 8 hours per day and 40 hours per week.

Training and Development

Training and development in Human Resource Management is the process of acquiring knowledge, skills and attitude that helps improve employee's job performance and enables future career growth. There is a difference between training and development. Training is a short term learning process where employees get an opportunity to develop skill, competency and knowledge as per the job requirement whereas development is a long term educational process which is concerned with the overall growth of the employees. Training focuses on present and it is job oriented whereas development focuses on future and it is career oriented. The objective of former is to improve the work performance of the employees but later is to prepare employees for future challenges.

Shiseido Company, Limited is a leader in the Japanese beauty industry with innovative skincare, suncare, and makeup products. It has a legacy of 150 years. Shiseido's commitment to work-life balance is evident in the high rate of employees returning from childcare leave (94.9%). Initiatives like "Welcome Back Seminars" and promoting paternity leave through employee stories aim to achieve a healthy work environment.

Tech Mahindra Limited is a leading global technology consulting company based in India that provides digital solutions and services to clients across industries. Tech Mahindra Limited has several policies that promote work-life balance, such as working from home, flexible work arrangements, parental care leave, in addition to paid time off. The company secured the third position on Vault's 2024 rankings for the best consulting firms in Asia-Pacific for work-life balance.

Maybank (KLSE: MAYBANK) stands as a financial services giant in Malaysia. It pioneered extended maternity leave in Malaysia, offering up to 365 days for female employees. This groundbreaking initiative set the stage for a continuous stream of policies and programs that champion family values and allow employees to integrate work with their personal lives.

To maintain healthy balance between paid work and personal life it is necessary to provide proper training and development not only for employees but also for employers so that employee can work in office without stress and pressure of employers. The major reason for poor relation between employee and employer is the factor "change", as employees don't want to change their way of work and employers expects maximum contribution and dedication of employees as per need of hours but no positive response from employees part causes ego clash and leads to disturbing the balance between work and personal life. From employee's perspective flexible working hours, work from home facility if the nature of the job permits healthy relations with superiors or employers, and cooperative attitude are the expectations of the employees and subordinates. Actually to bring balance between work and life one should be of very positive attitude towards others and self. Attitude of "Accepting the persons and the situations as they are" is very important. And this acceptance should be from within inside.

Gaur Gopal Das is a monk, lifestyle coach, motivational speaker and former electrical engineer. He is a member of the International Society for Krishna Consciousness. He emphasizes the importance of work-life balance by advocating for setting priorities, managing time effectively, and ensuring quality in both professional and personal spheres. He often stresses the need for boundaries, self-care, and maintaining meaningful relationships alongside career pursuits to achieve holistic happiness and success.

Sadhguru, a spiritual leader is the founder and head of the Isha Foundation, discusses work-life balance from a perspective of inner balance and fulfillment. He emphasizes that instead of dividing life into work and personal time, individuals should aim for an integrated approach where work becomes a joyful expression rather than a burden. Sadhguru suggests that aligning one's work with their passion and purpose can naturally create balance, as it reduces stress and enhances overall well-being. He also encourages practices such as mindfulness, meditation, and yoga to maintain inner harmony, which he believes is essential for achieving sustainable success and happiness in both professional and personal life.

Shree Shree Ravishankar, a spiritual leader and founder of the Art of Living Foundation, often emphasizes the importance of balance and harmony in all aspects of life, including work-life balance. His teachings on this topic typically include:

1. **Efficiency and Effectiveness**: He encourages individuals to prioritize tasks effectively and efficiently, ensuring that work is completed within reasonable time frames, allowing for adequate personal time and leisure.
2. **Purposeful Living**: He advocates for aligning work with one's deeper purpose and values. When work becomes meaningful and purpose-driven, it can contribute positively to overall life satisfaction and balance.
3. **Healthy Boundaries**: Establishing clear boundaries between work and personal life is crucial. This includes setting aside dedicated time for family, relationships, hobbies, and self-care to prevent burnout and maintain overall well-being.
4. **Stress Management**: Techniques such as breathing exercises, yoga, and relaxation techniques are often recommended by him to manage stress effectively, thereby supporting a balanced and fulfilling life.
5. **Integration of Inner Peace**: He suggests that true work-life balance begins with inner peace and contentment. By cultivating a peaceful mind through practices like meditation and mindfulness, individuals can better manage stress and achieve balance.

Overall, his teachings emphasize the integration of spiritual principles with practical approaches to achieve a harmonious work-life balance that nurtures both personal happiness and professional success. Samarthguru Siddharth Auliya is an Enlightened Master, Spiritual Scientist, Globalist, Poet, Author, PHD in Indian Schol Mines and former Lecturer in Indian School of Mines till 1975 and after that he worked as General Manager at Coal India Ltd till 2002. He did senior Executive Course in Mnanagement at Manchester Business School, UK in 1991. His 36 research papers and 4 books on Geology are his publications. For his book "Koyle

Ki Gaveshna" he was awarded Indira Gandhi Award in 1984. In 1992 he introduced an improvised and Indianised version of 'Quality Circle' aptly named 'Kaushal Chakra' in South Eastern Coalfields Ltd., thereby skyrocketing the company's profile and reducing accidents and strikes to a large extent. Ever since he was a child his spiritual journey started. He founded Indian Institute of Yoga and Management in 1988. He experienced enlightenment on 5 March 1997. He developed 28 levels of Spiritual courses and 42 wisdom courses which are very scientific for the seekers on the path of Enlightenment and beyond. The courses can be categorized as follows.

- Advanced Medidation Courses
- Blissful Living Courses
- Lifestyle and Wisdom Courses
- Corporate Courses
- Relationship, Parenting and Children Courses
- Health Courses

Training and development cell of every organization should organize workshops to bring balance in the life of their employees. Here are some suggestions and recommendations of few corporate courses designed by the organization for maintaing healthy balance between work and life of the peoples. These workshops are very scientific in nature and are trusted and endorsed by large corporate. Renowned companies like ONGC, Indian oil, Indian Railway, Coal India and many corporate organizations in India have already conducted these workshops and availing benefits of work life balance. IIYM believe in fostering a 'Blissful and Skillfull World' based on 'Work for Excellence' from the Shrimad Bhagwat Geeta. These courses are designed to provide a perfect balance of modern day management skills and ancient meditative practices to enhance productivity and efficiency. These courses aim to provide a holistic approach to skill building by emphasizing personal growth and development as well as professional skill building. The course bestows a new paradigm of living a life full of thrill and positivity which, in turn, instills high morale and boosts the working atmosphere of the organization.

S.No.	Name of workshop	Objective of the workshop	Duration of workshop	Output of the workshop
1.	Mind wellness workshop	This workshop is designed to provide a basic understanding of blissful living and its benefits in the workplace. It is suitable for all employees, regardless of their level of experience with mindfulness. To provide employees with the tools and techniques they need to reduce stress, increase focus, and improve communication in the workplace.	2 hours	Right mindfulness, stress free blissful, goal management, excellence, conflict management

S.No.	Name of workshop	Objective of the workshop	Duration of workshop	Output of the workshop
2.	Blissful living (Introduction to Right Mindfulness)	It focuses on developing the skill of mindfulness in order to achieve a greater sense of peace, bliss and fulfillment in one's life. To provide with a comprehensive understanding of Eightfold path of Gautam Buddha and how it can be applied in their daily lives. It covers various techniques and practices including modern day management skill and ancient meditative practices.	2 Days	Stress management, improving mental health and enhancing overall wellbeing
3.	Divine Living (Right Mindfulness level – 2)	Second level of right mindfulness, building upon the foundation established in blissful living. Here the participants will deepen their understanding and cultivate their spiritual selves learn to connect with their one's inner wisdom and developing a deeper sense of intuition.	3 days	Participants will learn to trust their intuition and use it as a guide in their daily lives. Learn to cultivate gratitude and find joy in every moment.
4.	The Awakened Living (Right Mindfulness level – 3)	It is a happiness course. It focuses on cultivating the ability to sustain mindfulness in all activities of daily life. Includes advanced mindfulness techniques to help individuals To provide a deeper understanding of underlying principles of mindfulness and their relationship to personal transformation and spiritual growth.	3 days	Develop a greater awareness of selves and to cultivate acceptance towards all.

Conclusion

Work life balance is not mere related to work and life; it is the positive state of mind. In its broader sense, is defined as a satisfactory level of involvement or fit between the multiple roles in a person's life. There is a need of proper training and development of employees and employers too for maintaining healthy balance between paid work and personal life. Organizing Workshops could help in improving the mental health and there to balancing their work and life of employees. These workshops of corporate courses include interactive sessions, experiential learning and real life scenarios to help participants to understand the nuances of corporate life. In addition, the course leverages the Noble Eightfold Path of Budha and incorprates meditation techniques, and focuses on Self Awareness and Right Mindfulness to help individuals to develop a deeper understanding of themselves and their place in the organization. Through this workshop, individuals can gain a competitive edge in the corporate world while also developing a deeper sense of purpose and fulfillment in their work. Output of the workshops will surely benefit the employer and employees to achieve a balance in their work and life. Actually a person cannot maintain a balance in life until and unless he/she practice meditation and yoga as a way of living. More than 80 thousand people from all over the world have attended the courses and have experienced total transformation from stress to happiness, failure to success and boredom to bliss.

References

Ameriya, S., & Singh, P. (2021). Work life balance of women employees in service sector during covid 19, (with special reference to indore city). In D. S. Jain, *Emerging issues in Law, Economic and Business* (pp. 17-24). RFI.

Gautam, I., & Jain, S. (2018). A study of work life balance : challenges and solutions. Research Gate.

https://www.oecdbetterlifeindex.org/topics/work-life-balance/. (n.d.). Retrieved from OECD Better Life Index.

https://finance.yahoo.com/news/20-asian-countries-best-life-140829294.html#:~:text=Although%20European%20countries%20dominate%20all,that%20promote%20work%2Dlife%20balance.

https://www.artofliving.org/in-en

https://isha.sadhguru.org/en/center/isha-yoga-center-coimbatore

https://samarthgurusiddharth.org/

CHAPTER 9

SUSTAINABLE RESOURCE MANAGEMENT

Dr. Diksha Khare

Asst.Prof. of Botany, Dept of Botany, Govt. N.PG. Science College, Raipur,(C.G)

Introduction

The term sustainable resources management refers to developing and implementing methods and solutions for using and protecting resources such as water, energy and raw materials as well as treating waste-streams and -waters. Sustainable resource management (SRM) can be defined as the application of sustainable practices by managing resources in a way that will benefit current and future generations. Simply put, it means managing resources in a way that ensures resources are not depleted to the point where it will impact future generations.

Why Sustainable Resource is Important?

Sustainable management of natural resources is important because it helps to use resources wisely without unnecessary use and without sacrificing on future generation's needs. - Reuse is better than recycling, since it takes both energy and resources to recycle materials. Moreover, there are few recyclable materials.

Aim of SRM

Sustainable resource management refers to integrating actions aimed at reducing negative environmental impacts and preserving natural capital throughout the life cycle of materials, considering economic efficiency and social equity.Once again, it is a question of acting in favor of ecology and the environment by limiting the waste of natural resources. Lately, many companies have been focusing more and more on green and blue energy.

Goals of sustainable development: 7GSD

- Structure of goals, targets and indicators. ...
- Goal 1: No Poverty. ...
- Goal 2: Zero hunger (No hunger) ...
- Goal 3: Good health and well-being. ...
- Goal 4: Quality education. ...
- Goal 5: Gender equality. ...
- Goal 6: Clean water and sanitation. ...
- Goal 7: Affordable and clean energy

Purpose of Sustainability

To pursue sustainability is to create and maintain the conditions under which humans and nature can exist in productive harmony to support present and future generations. Learn more about sustainability and how EPA incorporates it into its work in the National Research Council's report, Sustainability and primary goal of sustainability management:The objectives of sustainable development management: Join the objectives of the United Nations mentioned in their 2030 agenda. These include prosperity, peace, the well-being of people, partnerships and collaborations, and the conservation of the planet's resources the primary goal of sustainable living: Sustainable living is a lifestyle that attempts to reduce an individual's or society's environmental impact by making conscious decisions about everyday habits. Offsetting climate change, pollution, and depletion of resources are key goals.

Importance of Sustainable Living

With almost 8 billion people on the planet, living sustainably is more important than ever. Sustainable living is about making choices that reduce our impact on the environment. This includes everything from the food we eat to how we get around.

Sustainable living is important for several reasons. First, it helps to preserve our planet for future generations. With only one planet to call home, we need to do everything we can to protect it.

Second, sustainable living is about being more efficient with the resources we have, knowing that they are not unlimited. By being less reliant on fossil fuels, we can reduce our impact and slow down the process of climate change.

Third, sustainable living is vital for our health and the planet's health. Sustainable living encourages us to create healthier habits, such as cycling and eating organic food. These habits benefit not only our health but also the planet's health.

Finally, sustainable living is important because it is the morally right thing to do. We have a responsibility to take care of our planet and its resources. Sustainable living is the only effective way to curb climate change and preserve our planet.

Sustainable Living vs Zero-Waste Living

A lot of people, confuse sustainable living with zero-waste living. While they are similar, there are some key differences. Zero-waste living is about creating as little waste as possible. This means avoiding single-use items and choosing reusable options instead. It also means composting or recycling everything that you can. Sustainable living is about more than just waste. It is about making choices that are good for the environment. This includes everything from the food we eat to how we get around. Sustainable living is about creating a balance between meeting our needs and preserving our planet. For example, in terms of food, zero-waste living attempts to eliminate food waste by composting or recycling. On the other hand, sustainable living focuses on reducing our impact by eating locally and choosing organic options.

Regarding transportation, zero-waste living would focus on using public transport or carpooling to lessen emissions. Sustainable living, however, would also encourage walking or cycling.

Sustainable living is a much broader concept than zero-waste living. It is about making choices that are good for the environment in all our lives.

Going Green on a Larger Scale

While sustainable living is important for everyone, some people can make a bigger impact. This includes businesses and governments.

Businesses have a lot of power when it comes to sustainability. They can make choices that reduce their impact on the environment. This includes everything from the products they sell to how they package them. Governments also have a lot of power when it comes to sustainability. They can create policies that encourage sustainable living. For example, they could offer tax breaks for environmentally friendly businesses. They could also create laws that ban single-use plastics.

There is no one-size-fits-all solution when it comes to sustainability. But everyone can make a difference by doing their part. Whether you are an individual, a business, or a government, you can make choices that will help preserve our planet for future generations.

Final Thoughts

Sustainable living is important for several reasons. It helps preserve our planet and is vital for our health and the planet's health. We all are responsible for taking care of our planet and its resources. Living sustainably is the only effective way to curb climate change and preserve our planet.

If you want to learn more about sustainable living, plenty of resources are available. You can start by doing some research online or reading one of the many books that have been written on the subject. You can also talk to someone who is already living sustainably. They can offer advice and tips on how to make the transition.

The most important thing is to take action. Sustainability is not a trend or a fad. It is an essential part of preserving our planet for future generations. With climate change becoming an increasingly pressing problem, we all need to do our part to make a differenc

In 2015, all the countries in the United Nations adopted the 2030 Agenda for Sustainable Development. It sets out 17 Goals, which include 169 targets. These wide-ranging and ambitious Goals interconnect. Goal 3 is to ensure healthy lives and promote well-being for all at all ages. But it is also cross-cutting, so that progress in its implementation contributes to progress towards other Goals, and action on other Goals in turn contributes to attaining Goal 3. Most of the Goals also have some direct health targets. All of them have indicators by which progress can be measured.

Different about the Sustainable Development Goals:The 2030 Agenda and its Goals offer a comprehensive vision for sustainable development that:

- is global, rather than limited to "developing" countries as was the case with the Millennium Development Goals (MDGs);
- is based on values such as equity and respect for human rights;
- relies on approaches such as sustainable financing, scientific research and innovation, and monitoring and evaluation;
- requires a new way of working, involving intersectoral action by multiple stakeholders;
- aims to strengthen health systems towards universal health coverage (UHC).

Sustainable development and environmental priorities in India

The concept of SDM was first introduced b IUCN (international Union for conservation of nature and natural resources) in 1980. The Brundtland commission considered population control food secure and energy supply as critical component of sustainability. Economic development, social development and environmental protection are here independent and mutual reinforcing pillar of SD. According to the world resources institute 4 dimensions of SD are social economical environmental and technological. Economic development, which aimed at increasing the production of goods and services to meet the needs of a rising population, puts greater pressure on the environment. In the initial stages of development, the demand for environmental resources was less than that of supply. Now the world is faced with increased demand for environmental resources but their supply is limited due to overuse and misuse. Sustainable development aims at promoting the kind of development that minimizes environmental problems and meets the needs of the present generation without compromising the ability of the future generation to meet their own needs

How many SDGs are adopted by India

These efforts have strengthened the statistical systems and developed a monitoring framework covering all the 17 Goals and more than 100 indicators across the country. In India, financial inclusion has always been a priority, since 1969, when banks were nationalized, the strategy for addressing the banking needs of the poor has been biased toward providing credit, neglecting other aspects, such as building a deposit base, promoting a savings culture, or extending the payment network. Though over the last decade, India's financial inclusion agenda has seen an imperative move from an emphasis on credit to a more comprehensive approach toward financial services, specifically opening bank accounts and offering basic financial products like insurance, etc. This move has been partly driven by the objective to attain other public policy goals, such as replacing product subsidies with cash transfers, which requires people to have bank accounts for advancing the transfers.169 targets are part of a transformative agenda - the 2030 Agenda for Sustainable Development adopted by 193 Member States at the UN General Assembly Summit in September 2015, and which came into effect on 1 January 2016. At the core of this global agenda for 2030 is the principle of universality: 'Leave No One Behind'. Development in all its dimensions must include all people, everywhere, and should be built through the participation of everyone, especially the most vulnerable and marginalised. This comprehensive agenda recognises that it is no longer sufficient just to focus on economic growth, but on fairer and more equal societies, and a safer and more prosperous planet.

The status of India in SDGs:In this regard, the Government is implementing a number of targeted programmes such as Mahatma Gandhi National Rural Employment Guarantee Scheme (MGNREGS), Pradhan Mantri Awas Yojana-Gramin (PMAY-G), Pradhan Mantri Gram Sadak Yojana (PMGSY), Deendayal Antyodaya Yojana - National Rural Livelihoods Mission (DAY NRLM),

India's record in implementing Sustainable Development Goals

Since 2018, India has witnessed substantial progress in several key SDGs. Significant progress has been made in Goals 1 (No Poverty), 3 (Good Health and Well-being), 6 (Clean Water and Sanitation), 7 (Affordable and Clean Energy), 9 (Industry, Innovation and Infrastructure) and 11 (Sustainable Cities and Communities)

The Sustainable Development schemes implemented in India

Acting on its nationally-determined contributions, India has electrified 100% of its villages, reduced 38 million tonnes of CO2 emissions annually through energy efficient appliances, provided clean cooking fuel to 80 million poor households, and set a target to install 450GW of renewable energy and restore 26 million

Indian model of SDG

This Indian model of SDG localization is built around four foundational pillars: (i) creating institutional ownership; (ii) driving competition through collaborative competition; (iii) building capacities; (iv) adopting a whole-of-society approach

State is No 1 in SDG in India: Uttarakhand and Kerala have been named the top-performing states in NITI Aayog›s SDG India Index 2023-24, which measures the progress of states and union territories on social, economic, and environmental factors

References

United Nations General Assembly (1987) .*Report of the World Commission on Environment and Development: Our Common Future* Archived 31 March 2022 at the Wayback Machine. Transmitted to the General Assembly as an Annex to document A/42/427 – Development and International Co-operation: Environment.

Robert, Kates W.; Parris, Thomas M.; Leiserowitz, Anthony A. (2005). "What is Sustainable Development? Goals, Indicators, Values, and Practice". Environment: Science and Policy for Sustainable Development. 47 (3): 8–21. Bibcode:2005ESPSD..47c...8R. doi:10.1080/00139157.2005.10524444

"Sustainable Development". UNESCO. 3 August 2015. Retrieved 20 January 2022.

Environmental priorities in India and sustainable development Indian Science Congress Session, 73th, IN, 3-7 January 1986.

Collins, C. J., & Clark, K. D. (2003). Strategic human resource practices, top management team social networks, and frm performance: The role of human resource in creating organizational competitive advantage. Academy of Management Journal, 46(6), 740-51.

Fernandez, E., Junquera, B., & Ordiz, M. (2003). Organizational culture and human resources in nvironmentalissues. The International Journal of Human Resource Management, 14(4), 634-656.

Unnamalai, S.W. (2015), "Role of Indian bank's in Financial Inclusion Policy Through Pradhan Mantri Jan Dhan Yojana (PMJDY)", Sai Om Journal of Commerce & Management, Vol 2, Issue 12, December 2015.

Paramasivan. C and kamaraj, R. (2015), "Commercial Bank's Perforamce on Pradhan Mantri Jan Dhan Yojana", International Journal of Scientific Research And Education, Vol 3, Issue 6, June 2015, pp.3395-3603 .

Bharati Pajari., (2016), "Pradhan Mantri Jan Dhan Yojana (PMJDY): A Major Step towards Financial Inclusion in India", International Journal of Management Research and Business strategy, Vol 5, No 2, April 2016.

CHAPTER 10

INTERDISCIPLINARY APPROACHES TO CLIMATE CHANGE: THE CRUCIAL ROLE OF ENVIRONMENTAL PSYCHOLOGY

Dr. Anjana Purohit[1], Dr. Soumya Purohit[2]

[1]Department, Department of Psychology, Govt. Arts and Commerce Girls' College, Devendra Nagar, Raipur, Chhattisgarh.
[2]Peer Reviewer, Reserachhub.

Abstract

To capture the complete scope of the environmental challenges, Traditional single-disciplined approaches are not always sufficient. Multidisciplinary approaches are required to address the complex environmental problems of our time. Environmental issues like climate change, pollution, biodiversity loss, and deforestation are among the most critical challenges of our time. Among all these issues, climate change is the most multibranched challenge that goes beyond the limitations of traditional discipline, entailing a unified approach that includes economic, psychological, scientific, and behavioral perspectives. Environmental Psychology is a field that deals with the interchange between human behavior and the environment. It gives important insights into the psychological factors that motivate both individual and collective climate change action. This review paper investigates the critical role of environmental psychology in interdisciplinary strategies to climate change mitigation and adaptation.

Introduction

Climate Change

Climate change is one of the most urgent environmental challenges these times. Rising temperatures, sea-level rise, and extreme weather conditions affect human health, the earth's ecosystem and economic stability on a very high scale. As the impact of these issues intensifies, searching for effective solutions becomes highly significant. Addressing climate change necessitates a thorough understanding of its causes, consequences, and possible solutions (Pachauri & Reisinger, 2007, IPCC, 2021).

Traditional approaches to addressing these environmental challenges have often been sector-specific and reactive, focusing on isolated problems rather than considering the interconnected nature of environmental systems. Furthermore, traditional approaches have frequently prioritized short-term economic interests over long-term ecological considerations, exacerbating environmental problems and under-mining sustainable development goals (Clark & Dickson, 2003). Conventional approaches mostly focus on understanding the physical process of climate change, formulating policies based on economic models and developing technological solutions, which up to some extent can provide us with the solutions but they fail to address the complete complexity of the issue. The emergence of interdisciplinary methods seeks to bridge these gaps, integrating knowledge from multiple fields. This comprehensive approach acknowledges that a thorough grasp of social systems and human behaviour is just as important for effective climate action as technological and legislative solutions.

Environmental Psychology

Environmental psychology, primarily climate psychology, investigates the psychological aspects that influence how individuals and groups perceive, respond to, and interact with climate change (Jasonoff, 2004). Understanding these aspects are critical for developing interventions that encourage sustainable behaviours, increase public support for climate policy, and foster resilience to climate effects. Psychological considerations can help with a variety of climate change mitigation attempts, including developing economic incentives and building urban environments that promote sustainable living (Frumkin, Frank, & Jackson, 2004). Integrating psychological theories and methods into interdisciplinary approaches allows us to design more effective ways for cutting emissions, responding to climate impacts, and establishing a more sustainable future (Ostrom, 2009).

Recent studies demonstrate the significance of psychological treatments in climate change communication. For example, stressing the co-benefits of climate action, such as improved health and cost savings, can boost public participation and support for climate policies (van der Linden et al., 2015). Furthermore, framing climate change in a way that speaks to people's values and identities has been proven to boost the likelihood of pro-environmental behaviour (Whitmarsh et al., 2013;

Corner et al., 2014).). Behavioural economics, for example, employs psychological concepts to create policies that encourage people to make ecologically responsible choices, such as taking public transportation instead of driving (Gifford, 2011; Swim et al., 2011). Similarly, green space-based urban planning can promote mental health and increase community support for sustainability projects (Maffi, 2005).

This review will look at how environmental psychology fits into interdisciplinary climate change approaches, the role of human behavior in affecting climate outcomes, and the potential for environmental psychology to contribute to better and more sustainable alternatives. It will examine how psychological insights can help shape policies and practices in behavioral economics, urban planning, and communication. The subsequent sections discuss the practical uses of these interdisciplinary tactics through extensive analysis and examining the challenges and opportunities they bring. And exploring deeper into how psychological insights can potentially be applied to various disciplines to encourage sustainable behaviour and more effectively combat climate change.

Understanding Climate Change and the behaviour of People

Climate change is defined as long-term changes in temperature, precipitation patterns, and other atmospheric variables on Earth. These changes are mostly caused by human activities, particularly the use of fossil fuels, deforestation, and industrial operations, which raise atmospheric concentrations of greenhouse gases (GHGs) such as carbon dioxide (CO_2) (IPCC, 2021). Climate change has profound consequences, including increased frequency and severity of weather events, increasing sea levels, shifting agricultural zones, and biodiversity loss (Stern, 2007; Pachauri & Reisinger, 2007).

The behaviour of humans plays a significant part in both causing and alleviating climate change. Energy usage, transportation choices, food habits, and waste management practices all have a direct impact on GHG emissions (Gifford 2011; Abrahamse et al., 2005). Understanding the psychological causes of such patterns is critical in creating interventions that encourage more sustainable practices. Environmental psychology investigates these drivers, including attitudes, beliefs, values, social norms, and habits that influence environmental behaviour (Stern, 2000; Whitmarsh et al., 2013).

Key Behavioural Factors Contributing to Climate Change

Energy Consumption

The use of energy in homes, industries, and transportation contributes significantly to greenhouse gas emissions. Behaviours such as excessive power consumption, reliance on non-renewable energy sources, and poor energy practices enhance the issue (Gifford, 2011; Abrahamse et al., 2005). According to studies, adopting energy-saving practices such as turning off lights when not in use and utilizing energy-efficient appliances can considerably cut emissions (Allcott and Mullainathan, 2010).

Choice of Transportation

Personal transportation practices, such as the preference for driving over public transportation, cycling, or walking, accounted for a significant number of harmful emissions. The transportation sector contributes significantly to global emissions, particularly using gasoline and diesel in automobiles (Ewing & Cervero, 2010). Studies show that initiatives that promote public transit, carpooling, and electric vehicle usage can help to reduce emissions (Cervero & Murakami, 2010).

Dietary Habits

Dietary preferences, particularly those including meat and dairy products, have a considerable environmental impact. Livestock production is a substantial producer of methane and nitrous oxide, both powerful greenhouse gases (de Boer et al., 2014). Plant-based diets and reduced meat intake can help to reduce agricultural greenhouse gas emissions (Lea et al., 2006).

Waste Management

Improper waste management methods, such as inadequate recycling and the overuse of single-use plastics, contribute to environmental damage and hazardous emissions from landfills (Jackson 2005). Recycling, composting, and reducing plastic use are critical steps toward more sustainable waste management techniques (Gifford, 2014).

Psychological Dimensions of Climate Change

Perception and Awareness

Understanding the psychological dimensions of environmental behavior is critical for establishing successful ways to promote sustainable practices and mitigating climate change. Environmental psychology provides useful information on the elements that influence how people perceive, respond to, and deal with environmental challenges. This section investigates the psychological drivers of environmental behavior, such as attitudes and beliefs, values and standards, habits and routines, and perceived control and efficacy.

Attitudes and perceptions about the environment have a substantial influence on behavior. People with strong pro-environmental views are more inclined to recycle, conserve energy, and support environmental regulations (Ajzen, 1991). People who closely identify with the environment are more likely to engage in environmentally friendly practices. Environmental identity represents how related people view themselves to the natural world and how important environmental issues are to their self-concept (Clayton, 2003).

When people have contradictory attitudes or ideas, they may suffer cognitive dissonance, which can encourage behaviour change to alleviate the discomfort created by these discrepancies. Individuals who believe in the necessity of environmental conservation but do not recycle can change their actions to align with their ideas (Festinger, 1957).

Values and Social Norms

Values and social norms have an important role in shaping environmental behavior. Values are long-lasting beliefs about what is important in life, whereas social norms are unwritten laws that govern behavior in a specific culture. Schwartz's value theory specifies numerous values that are especially relevant to environmental behavior, including universalism and compassion. People who emphasize these principles are more likely to engage in environmentally friendly practices (Schwartz, 1992). Descriptive norms are perceptions of how other people behave. Individuals are more inclined to engage in pro-environmental acts if they see their peers doing so (Cialdini, 2003).

Habits and Routines

Habits and routines are automatic behaviors that need little conscious thought. These habits are often tough to alter, but understanding their genesis and disruption can aid in the development of successful treatments to promote long-term behaviour. Habit formation involves repetition and reinforcement. Recycling or turning off lights are examples of environmental activities that can become habitual when undertaken repeatedly in response to certain cues (Verplanken & Orbell, 2003). Changing current behaviours can be difficult, but interventions like reminders, feedback, and incentives can assist. For example, installing recycling containers in handy locations can encourage people to recycle more frequently (Neal, Wood, & Quinn, 2006).

Perceived behavioural control and self-efficacy pertain to people's views about their ability to conduct an activity and influence the outcome. These views have a major impact on the chance of engaging in pro-environmental practices. Higher perceived control increases the likelihood of engaging in the behaviour (Ajzen, 1991). Increasing self-efficacy might encourage people to engage in pro-environmental actions such as limiting energy usage or campaigning for environmental policies (Bandura, 1997).

Motivation

Motivation is also an important driver of the behaviour, both intrinsic and extrinsic. Individuals who are intrinsically driven to engage in pro-environmental acts frequently do so because they find the actions rewarding and consistent with their values (Ryan & Deci, 2000). Policies that offer financial incentives for energy efficiency or punish excessive emissions can employ extrinsic motivation to encourage sustainable behavior (Deci, Koestner, & Ryan, 1999).

Emotion, Emotional Responses, and Climate Anxiety

Emotions play an important influence in shaping environmental behaviour. Emotional responses to environmental challenges can either drive or discourage action. Fear-based messages about the effects of climate change can often backfire by making people feel overwhelmed or despairing. However, when accompanied with practical answers, fear might encourage protective behaviours (O'Neill &

Nicholson-Cole, 2009). Positive emotions like hope and optimism can motivate people to act because they make them feel empowered and capable of making a difference. Campaigns that promote success stories and accomplishments can help to foster happy feelings (Ojala, 2012).

Understanding the psychological foundations of environmental behaviour is critical for designing successful treatments and policies that promote sustainability. By addressing attitudes and beliefs, values and norms, habits and routines, perceived control and efficacy, motivating variables, and emotional responses, we can develop tactics that promote pro-environmental behaviour and contribute to climate change mitigation.

Interdisciplinary Approaches Involving Environmental Psychology: Case Studies and Practical Applications

Integrating Behavioural Economics and Climate Policy

Behavioral economics, which explores how psychological, cognitive, emotional, cultural, and social aspects influence economic decisions, can help policymakers build effective climate policies. Traditional economic models usually presume rational conduct, whereas behavioral economics recognizes that people make decisions that are not reasonable. This insight can be used to develop rules that better reflect actual human behavior. For example, "nudging" tactics, which gradually encourage decisions without limiting options, have been effective in encouraging pro-environmental behavior. These techniques can include making green energy the standard choice, using social comparisons to lower energy usage, and structuring messaging to emphasize immediate personal benefits, such as cost savings (Thaler & Sunstein, 2008; Allcott & Mullainathan, 2010).

Case Study: Opower Program. The Opower initiative, which sends households energy reports that comparing their usage to that of their neighbors, has dramatically cut energy consumption. This behavioral intervention has reduced energy use significantly by leveraging social norms and delivering unambiguous feedback (Allcott, 2011).

Urban Planning and Sustainable Development

Urban planning is vital to developing sustainable cities and communities. Environmental psychology investigates how the design and structure of urban environments affect human behavior and well-being. Integrating these ideas into urban planning can result in cities that are not only environmentally sustainable but also improve residents' health and well-being. Green infrastructure, including parks, green roofs, and urban forests, has been found to improve air quality, reduce urban heat islands, and boost mental health. Furthermore, creating walkable areas and encouraging active transportation can reduce reliance on fossil fuels and lower greenhouse gas emissions (Frumkin, 2001; Gill et al., 2007).

Case Study- Copenhagen's Bicycle Infrastructure Copenhagen is well known for its extensive bicycle infrastructure, which promotes cycling as a key means of transportation. This design not only lowers carbon emissions, but it also encourages physical activity and enhances public health. The city's strategy combines urban planning with psychological insights into human behavior and preferences (Pucher and Buehler, 2008).

Educational and Communication Strategies

Effective communication is critical for enhancing awareness of climate change and driving behavior change. Environmental psychology provides ideas for developing communication campaigns that appeal to a wide range of groups. Emphasizing the co-benefits of climate action, such as increased health and economic savings, can boost public support. Furthermore, framing climate change in ways that reflect people's values and identities enhances the likelihood of pro-environmental behavior (Maibach et al., 2010; van der Linden et al., 2015).

Case Study: Climate Change Awareness Programs. Interactive and immersive learning strategies have been shown to increase student comprehension and concern about climate change. Programs that include children in hands-on activities and critical thinking exercises help them connect with the topic on a deeper level, creating long-term effects.

Health and Well-being

Environmental psychology coincides with public health, emphasizing how climate change and environmental degradation affect both mental and physical health. Green urban planning and active transportation can help to reduce stress, improve mental health, and raise physical activity levels. Addressing climate change from a public health standpoint can also inspire action by making the problem more current and personal for individuals (Frumkin et al., 2004).

Case Study on Heatwaves and Urban Planning. According to research, urban areas with greater green space and tree cover can better withstand heat waves. These green infrastructures can lower temperatures, lowering the risk of heat-related illnesses and deaths. Environmental psychology can help cities become more resilient to climate impacts while also improving citizens' well-being (Stone et al., 2010).

Community-Based Approaches

Community-based methods capitalize on local expertise and promote shared responsibility for climate action. Environmental psychology sheds light on the social dynamics that generate community engagement and empowerment. Participatory planning, in which community people actively participate in decision-making processes, is frequently used in successful programs, leading to increased acceptability and commitment to sustainability projects (Kollmuss & Agyeman, 2002).

Case Study: Transitional Towns. The Transition Town movement encourages communities to generate local solutions to minimize carbon footprints and strengthen resilience to climate change. These programs frequently include community gardens, local energy projects, and skill-sharing seminars to foster social cohesiveness and collective efficacy (Hopkins, 2008).

Conclusion

Environmental Psychology offers valuable perspectives on how individuals and communities perceive and respond to climate change. Attitudes and beliefs, values and social norms, habits and routines, perceived control and efficacy, motivation, and emotional responses are all important psychological dimensions that influence environmental behavior. Moreover, integrating environmental psychology with other fields, such as behavioral economics, urban planning, public health, and community engagement, has the potential to greatly improve climate change mitigation and adaptation measures. Understanding and addressing the psychological causes of human behavior allows us to build more comprehensive and successful approaches to preventing climate change. Variety of interdisciplinary approaches that draw on these psychological insights can boost public participation, encourage pro-environmental behavior, and aid in the creation of resilient and sustainable communities. Future research and practice should focus on these multidisciplinary connections, encouraging cross-disciplinary collaboration to create a sustainable and resilient future.

References

Ajzen, I. (1991). The theory of planned behavior. *Organizational Behavior and Human Decision Processes, 50*(2), 179-211.

Allcott, H. (2011). Social norms and energy conservation. Journal of Public Economics, 95(9-10), 1082-1095.

Bandura, A. (1997). Self-efficacy: The exercise of control. Freeman.

Cialdini, R. B. (2003). Crafting normative messages to protect the environment. Current Directions in Psychological Science, 12(4), 105-109.

Cialdini, R. B., Reno, R. R., & Kallgren, C. A. (1990). A focus theory of normative conduct: Recycling the concept of norms to reduce littering in public places. Journal of Personality and Social Psychology, 58(6), 1015-1026.

Clayton, S. (2003). Environmental identity: A conceptual and an operational definition. In S. Clayton & S. Opotow (Eds.), Identity and the natural environment: The psychological significance of nature (pp. 45-65). MIT Press.

Deci, E. L., Koestner, R., & Ryan, R. M. (1999). A meta-analytic review of experiments examining the effects of extrinsic rewards on intrinsic motivation. Psychological Bulletin, 125(6), 627-668.

Festinger, L. (1957). A Theory of Cognitive Dissonance. Stanford University Press.

Frumkin, H. (2001). Beyond toxicity: Human health and the natural environment. American Journal of Preventive Medicine, 20(3), 234-240.

Frumkin, H., Frank, L., & Jackson, R. (2004). Urban Sprawl and Public Health: Designing, Planning, and Building for Healthy Communities. Island Press.

Hopkins, R. (2008). The Transition Handbook: From Oil Dependency to Local Resilience. Chelsea Green Publishing.

Kollmuss, A., & Agyeman, J. (2002). Mind the gap: Why do people act environmentally and what are the barriers to pro-environmental behavior? Environmental Education Research, 8(3), 239-260.

Maibach, E. W., Nisbet, M., Baldwin, P., Akerlof, K., & Diao, G. (2010). Reframing climate change as a public health issue: An exploratory study of public reactions. BMC Public Health, 10(1), 299.

Monroe, M. C., Plate, R. R., Oxarart, A., Bowers, A., & Chaves, W. A. (2019). Identifying effective climate change education strategies: A systematic review of the research. Environmental Education Research, 25(6), 791-812.

Neal, D. T., Wood, W., & Quinn, J. M. (2006). Habits—A repeat performance. Current Directions in Psychological Science, 15(4), 198-202.

Ojala, M. (2012). Hope and climate change: The importance of hope for environmental engagement among young people. Environmental Education Research, 18(5), 625-642.

O'Neill, S., & Nicholson-Cole, S. (2009). "Fear won't do it": Promoting positive engagement with climate change through visual and iconic representations. Science Communication, 30(3), 355-379.

Pucher, J., & Buehler, R. (2008). Making cycling irresistible: Lessons from the Netherlands, Denmark and Germany. Transport Reviews, 28(4), 495-528.

Ryan, R. M., & Deci, E. L. (2000). Intrinsic and extrinsic motivations: Classic definitions and new directions. Contemporary Educational Psychology, 25(1), 54-67.

Schwartz, S. H. (1992). Universals in the content and structure of values: Theoretical advances and empirical tests in 20 countries. Advances in Experimental Social Psychology, 25, 1-65.

Stone, B., Hess, J. J., & Frumkin, H. (2010). Urban form and extreme heat events: Are sprawling cities more vulnerable to climate change than compact cities? Environmental Health Perspectives, 118(10), 1425-1428.

Thaler, R. H., & Sunstein, C. R. (2008). Nudge: Improving Decisions About Health, Wealth, and Happiness. Yale University Press.

Van der Linden, S., Maibach, E., & Leiserowitz, A. (2015). Improving public engagement with climate change: Five "best practice" insights from psychological science. Perspectives on Psychological Science, 10(6), 758-763.

Verplanken, B., & Orbell, S. (2003). Reflections on past behavior: A self-report index of habit strength. Journal of Applied Social Psychology, 33(6), 1313-1330.

CHAPTER 11

EXPLORING THE NUTRITIONAL & HEALTH BENEFITS OF RED RICE WITH SPECIAL REFERENCE TO CHHATTISARH: A COMPREHENSIVE REVIEW

Bhabita Mandavi[1], Dr. Deepti Chauhan[2], Manisha Nayak[3]

[1,3]Assistant Professor, Department of Botany & Chemistry, Rani Rashmi Devi Singh Govt. College, Khairagarh, (C.G.)

[2]Assistant Professor, Department of Botany, Bhilai Mahila Mahavidyalaya, Hospital Sector, Bhilai (C.G.)

**Email: bhabitam10@gmail.com*

Abstract

Rice is the vital food grain that is mostly available in Asian countries. It is traditionally grown in Asian countries and is rich in protein, carbohydrates, and energy for human diets. Rice comes in a range of colors, including purple, red, brown, black, and yellow, which is the most popular. In the past, people preferred to eat colored rice because of its flavor and therapeutic properties. However, the coloured rice is remained unrecognized due to the commercialisation of agriculture and new plant introduction. If we talk about Chhattisgarh the red rice is commonly known as **"PASHAR CHAWAL/BHAT "**it is the primitive and traditional variety of rice which is popular in this region for its nutritional benefits and distinct taste. Red rice does have the richness and goodness of fibres, iron, vitamins and antioxidants it has various health benefits which makes it a good choice for regular or daily consumption. The name **"PASHAR CHAWAL"** indicates its provincial significance and cultural vitality in Chhattisgarh s cuisine but it has become the victim of partiality and it has not just the nutritious food choice but also a symbol of cultural identity and also a heritages significance with environmental productivity and positive concern for the regions where it is grown. Rice is grown by using suitable and sustainable farming practices although it is a water intensive crop hence that minimizes or reduces the use of chemical fertilizers and pesticides this approach often protects or conserves the environment and make it free from harmful essence of chemicals and prevents biomagnification. Red rice e belongs to the family of grass it is a remarkable gift of mother nature hence in this article the significance and importance of rice variety has been depicted.

Keywords: Nutritional, Traditional, Religion, Medicinal, Red Rice, Pashar Chawal, C.G.

Introduction

Grown in nearly all of India's 26 states and 7 territories, rice (Oryza sativa L.) is a key crop of the subcontinental region and the primary food crop for the country's half of a billion people. It is grown in a variety of agricultural climates. Since ancient times, eastern Indian farmers have been growing traditional landrace rice. (Dikshit et.al.,2013). The importance of rice, aside from being a regular food source, is also a source of income for peasants and the everyday requirement of millions of people. Rice can also be used as a vital commodity because its presence cannot be replaced by other commodities or edible goods and must be in the sufficient quantities (Setiowati et.al.,2019).

Red rice is defined as having pigments and a layer of reddish bran.it can be seen in wild weedy and cultivated types. China, India, Korea, Indonesia, Sri Lanka, and the United States have all grown red rice. West Bengal is the state that produces the most rice in India, followed by Madhya Pradesh, Uttar Pradesh, Punjab, Kerala, Orissa, and Chhattisgarh. In the developing world, rice (Oryza sativa L.) is the most important crop in terms of both overall output and the number of people who rely on it as a staple diet (Kaur& Iraj 2023). The major rice producing country is China followed by India, Bangladesh, and Indonesia. In areas like Chhattisgarh, where it is referred to locally as "Pashar chawal," red rice is very well-known in India. There is cultural and regional significance to this type of rice.

Chhattisgarh, is the land which represents the tribal culture where 70% of the population having agriculture as occupation (subsistence agriculture). Due to its vast production of rice varieties and its rich agricultural landscape and over 20,000 rice varieties have been recorded in the province, therefore, conventionally Chhattisgarh is famous as the "**Rice Bowl of India**". These are the products of millennia of rice cultivation by native societies, resulting from selection and adaptation to a range of water, soil, and microecosystem conditions, as well as the presence of predators like mice, insects, and cattle. Different land races and kinds were evolved by farmers over generations to suit varying geoclimatic and socioeconomic situations. Naturally occurring rice variants are abundant and represent an untapped gene pool. One distinctive aspect of many communities in Chhattisgarh is their well-established traditional knowledge systems of nutrition and self-healing, which are founded on their understanding of the Bio wealth. (Dwivedi et.al.,2016).

Origin & History

The red rice is generally occurred in nature they are obtained from the forest area. Primitively it is believed that the origin of red rice takes place in Asian countries such as - India, Bhutan and Sri Lanka followed by it is also considered that it is native of Africa too. *Oryza glaberrina* species of red rice found there in the past 3000 years ago. The important information regarding the origin of red rice also found in the Vedas and Puranas. The three most well-known founders of ancient Ayurveda in India are Charak (700 BC), Vagbhata (700 AD) and Susruta (400 BC) (Ahuja et.al.,2013). The red husk grain, a form of red rice with beneficial health effects and potent medicinal qualities, is discussed by Charak in the Charak Samhita. The Vishnu Purana (200 AD), Garuda Purana and Agni Purana (900 AD), all highlight how important it is for wild *nevara* (red-grained rice) to sustain the tridosha (vata, pitta, and kapha) and how it can also be used to counteract the effects of subsequent tridosha. (Saxena, 2014).

Characteristics of Red Rice

Red rice is those that have a layer of red bran on them. Even after a high degree of grinding, there is still a hint of red, even though it is limited to the bran layer. The bran varies in hue from pale to deep crimson. The bran layer has antioxidant properties and is composed of polyphenols and anthocyanins. Both red and white rices have a similar, white inside. Red rices have a 2-4 times higher zinc and iron content than white rice (Hettaet.al.,2022). The length and shape of red rice grains can vary, ranging from long and thin to shorter, nearly spherical grains. Red rice has a unique flavour with a little earthy undertone that is slightly nutty. After cooking, red rice often has a slightly aromatic aroma and a firm, chewy texture. Its entire bran layer and increased fiber content make it more flavourful and satisfying than white rice.

Photo plate 1. Grains of red rice (Pashar chawal)

Habitat & Ecology

Red rice grows in proper sunlight and is found in swampy and marshy areas, at the fringes of lakes or ponds often with underlying substratum river or stream sides. It is present in and at the edges of rice fields and in irrigation canals and in permanently wet or seasonally dry areas and water up to 4 m deep, but usually in 1m or less. The Red rice prefers black cotton soils, Kalahari sand, and alluvium and grow in a brackish water swamp. It has been found growing between sea level and 2000 m altitude.

Photo plate 2. Red rice (Pasar chawal)

Types of red rice

Red rice comes in a variety of forms, including cultivated, wild, and light- or dark-colored husks filled with red kernels.

Wild red rices: Of the different kinds of wild rice, India is home to *Oryza rufipogon, O. nivara Oryza granulata and O. officinalis. Red-grained O. nivara* and *O. rufipogon* are used as food and medicinal. The Sanskrit root niv, which meaning caloric or feeding, is whence the word *"nivara"* originates.

Weedy rices: Considered weedy are plants that are off-type and have red pericarps or seeds. Scientists think that hybridization between wild and cultivated rices is how these rices arose. The poor in India are the ones who eat weedy rice. Red weedy rice is currently not preferred in high-quality rice and is distinguished by dealers from low-grade rice. Certain weedy rices are left behind during harvesting in some places, and they are subsequently gathered for consumption or other uses.

Cultivated red rices: Earlier the introduction of highly productive white rice varieties, the red rice was almost important crop in all Asian Countries. These nations included Philippines, Sri Lanka, Korea, India, Japan, and China. In India, red rice was widely consumed in the Northeast, West, and South, as well as in the country's hilly regions. Many red-grain cultivars were grown in regions with unfavourable weather circumstances, including drought, sandy soils, deep water, cold temperatures, and salinity, including Kerala, Tamil Nadu, Karnataka, Bihar, Orissa, Bengal, Madhya Pradesh, and the northeastern states (Ahuja et.al.,2013).

Nutrients

Rich in dietary Fiber: The Red rice contains fiber in abundance which makes the food suitable for digestion.

Minerals and Vitamins: Red rice has the rich content of minerals such as-zinc, magnesium, iron and it is a good source of vitamin B, which is important for the functioning of biological activities.

Antioxidants: In red rice anthocyanin pigment found in abundance which is the reason behind its red coloured appearance due to this pigment the antioxidant properties are found in the red rice which helps to eradicate the harmful free radicles from the body.

Religion& Culture

Two rituals called Rishi Panchami, and Hal Shashti, are observed in India, Maharashtra, in August and September. Unlike produced crops with the help of animals, in these occasions instead uses of wild rice (*O. nivara*) and self-grown plants are expected from fasting women. (Ahuja et.al.,2013). Red rice has great cultural and religious significance in Chhattisgarh, particularly during the "Halshashti festival "Pashar chawal" is the name given to red rice in Chhattisgarh. The majority of people who celebrate this event, also known as "Har Chhath" or "Lalhi Chhath," are women, particularly moms living in rural Chhattisgarh. The event honours "Goddess Shashti," who is revered for her protection and general well-being over.

Significance of red rice in Halshashti

Health and Nourishment: Red rice helps to regulate and maintain the blood sugar level it contains low glycemic index which is responsible for the lowering of blood sugar level and the high fiber contain of red rice keeps the gut healthy.

Cultural offerings: In Hindu ceremonies rice is an emblem of purity fertility and Prosperity. In the festival of Halshasti women Cooke the authentic and cultural dishes from red rice to offer the motherly Goddess Shasthi which is often considered as a ritual to seek blessings for every individual and children.

Cultural Heritage: Being a traditional crop of Chhattisgarh, red rice is closely associated with the state's rituals and customs. It is also frequently used in a variety of folk snacks and dishes during Halshasti. The state has made significant efforts to preserve its agricultural diversity and customs by using sustainably produced indigenous rice varieties, such as red rice, as well as community customs where mothers used to bless their children in a special way by having them fast and dipping a small piece of cotton cloth into a mixture of holy water (Gangajal) and haldi (turmeric). The mother would then apply the mixture to her child's back, or trunk.

Community and Ceremony: These festivals also act as a cultural mediator bringing together people from different backgrounds and fostering sense of unity and belongings, they also promote community members to come together engage in shared experiences and build relationships in a better manner by improving soil relations.

Photoplate 3.1. Halshashti festival

Photoplate 3.2. Halshashti festival

Medicinal Impact

Nivara rice has the unique therapeutic ability to correct any tridosha imbalances, according to ancient Indian Ayurvedic literature. This rice is used extensively in Ayurvedic medicine because of its extraordinary ability to balance blood pressure, strengthen, rejuvenate, and energise the body, remove harmful metabolites, and prevent skin conditions and early aging. In several regions of India, it is also consumed on the customary fasting days. (Ahuja et.al.,2013). These days, experts even recommend it as a medication substitute for treating minor symptoms of high cholesterol. Red rice is high in iron and vitamins, which work together to support the body's synthesis of red blood cells, or RBCs, which are thought to be vital for maintaining healthy skin. Rice's antioxidants may aid in the battle against free radicals, which shield our skin. Due to its high manganese and antioxidant content, red rice is strongly advised for diabetic patients.

Cultivation

In practically every Asian nation that grows rice, including Philippines, Sri Lanka, Korea, China, India, and Japan, red rices played a significant role in the rice industry prior to the advent of high-yielding varieties, which are primarily white. A review of the literature indicates that red rices make up around 20% of the germplasm gathered in China. In India, the mountainous regions of the Northeast and West, as well as the, were the main producers of red rice. Red rice cultivars demonstrated a high level of resistance to unfavorable conditions, including mountainous terrain, deep water, and infertile soils. Many red-grain cultivars were grown in regions with unfavorable conditions such as drought, deep water, salinity, freezing temperatures, sandy soils, and in Tamil Nadu, Karnataka, Kerala, Orissa, Bihar, Madhya Pradesh, Bengal and the northeastern states. Red rices are very diverse, just as white rices. They can be both short- and long-grained, late-maturing, fragrant or non-aromatic, and glutinous or non-glutinous. But the majority of red rices have coarse grains. Himalayan red rice is one variety of long-grained red rice that has gained popularity in the international market (Hetta et.al.,2022).

Types of red rice cultivated in Chhattisgarh

The red rice traditionally consumed during the Halshashti festival in Chhattisgarh is primarily grown in specific regions of the state known for their individual agricultural practices. These regions include: Rajnandgaon, Bastar and Kanker. The climate in these regions depicted by adequate rainfall and fertile soil, is ideal for rice cultivation, and rich organic matter in the soil contributes to the robust flavour and high nutritional content of the red rice. The use of traditional and organic farming practices in these areas helps maintain the quality of the red rice, making extremely suitable for religious and cultural festivals like Halshashti. Chhattisgarh known for its rich agricultural diversity, cultivates several varieties of red rice. The region is home to multiple traditional and native rice varieties, many of which have unique characteristics and cultural significance. While the exact number can very due to local classification and naming conservations, here are some of the well-known types of red rice cultivated in Chhattisgarh: Lal Bhat, kattuyanam, Jhilli, Jharkhani Ramdhan, Indrakant or Indrani Dhan and Desi red rice.

Result & Discussion

In this article, we have studied about the variety of red rice found in Chhattisgarh, especially Pashar chawal. During the study, we have made an initial attempt to collect traditional knowledge of red rice, especially native red rice of Chhattisgarh. Along with this, information was also gathered about the nutrients, therapeutic importance, and special qualities of many landraces from farmers and tribal people. Red rice is valued for its potential medical benefits in addition to its nutritional value. Among its applications are: Because red rice contains anthocyanins, which function as antioxidants, it can aid the body fight off dangerous free radicals. Because they contain the least amount of fat, they may assist enhance cardiovascular health by lowering cholesterol and promoting heart health in general. Red rice's high fiber content can facilitate digestion and ward off constipation. Compared to white rice, red rice has a lower glycaemic index, which can aid with blood sugar regulation. Red rice may include a chemical with anti-inflammatory qualities that can help the body's natural inflammatory response.

Conclusion

Red rice is indeed a unique gift of nature. Pashar chawal or red rice holds significant agricultural, nutritional, economic, and cultural value in Chhattisgarh. its rich nourishing profile and cultural importance make it a vital part of the state s agricultural landscape. The cultivation and consumption of pasar chawal align with the goals of sustainable development, particularly in developing agrobiodiversity. By preserving traditional varieties like Pashar chawal, communities contribute to the resilience of agricultural systems and reduce dependency on monoculture practices, which can lead to biodiversity loss. However, to sustain and inflate its production and marketability, efforts are needed to address the challenges posted by modern agricultural practices and to improve market penetration. Through supportive policies and increased apprehension and awareness, Pashar chawal can continue to proliferate as both cultural staple and an economic resource.

References

Ahuja, U., Ahuja, S. C., Chaudhary, N., & Thakrar, R. (2007). Red rices–past, present and future. *Asian Agri-History, 11*(4), 291-304.

Carcea, M. (2021). Value of wholegrain rice in a healthy human nutrition. Agriculture, 11(8), 720.

Chaudhari, P. R., Tamrakar, N., Singh, L., Tandon, A., & Sharma, D. (2018). Rice nutritional and medicinal properties: A review article. Journal of Pharmacognosy and Phytochemistry, 7(2), 150-156.

Chen, Q. M., Fu, M. R., Yue, F. L., & Cheng, Y. Y. (2015). Effect of superfine grinding on physicochemical properties, antioxidant activity and phenolic content of red rice (Oryza sativa L.). Food and Nutrition Sciences, 6(14), 1277-1284.

Ciulu, M., Cádiz-Gurrea, M. D. L. L., & Segura-Carretero, A. (2018). Extraction and analysis of phenolic compounds in rice: a review. Molecules, 23(11), 2890.

Dikshit, N., Das, A. B., Sivaraj, N., & Kar, M. K. (2013). Phenotypic diversity for agromorphological traits in 105 landraces of rice (Oryza sativa L.) from Santhal Parganas, Jharkhand, India. Proceedings of the National Academy of Sciences, India Section B: Biological Sciences, 83, 291-304.

Ding, J., Ulanov, A. V., Dong, M., Yang, T., Nemzer, B. V., Xiong, S., ... & Feng, H. (2018). Enhancement of gama-aminobutyric acid (GABA) and other health-related metabolites in germinated red rice (Oryza sativa L.) by ultrasonication. Ultrasonics sonochemistry, 40, 791-797.

Duménil, G., Montminy, T. P., Tang, M., & Isberg, R. R. (2004). IcmR-regulated membrane insertion and efflux by the Legionella pneumophila IcmQ protein. Journal of Biological Chemistry, 279(6), 4686-4695.

Durand-Morat, A., Nalley, L. L., & Thoma, G. (2018). The implications of red rice on food security. Global Food Security, 18, 62-75.

Dwivedi, A. P., Chand, P., Mishra, A., Singh, S. R. K., & Athare, T. (2016). Identification of traditional rice varieties in Chhattisgarh: An institutional arrangement. Ecology, Environ. Conser, 23, 313-320.

Gealy, D. R., Mitten, D. H., & Rutger, J. N. (2003). Gene flow between red rice (Oryza sativa) and herbicide-resistant rice (O. sativa): implications for weed management. Weed technology, 17(3), 627-645.

Hetta G., Babal B., Sharma G.D, Kumar R. and Rana S.S. (2022). Organic By Default: Hill State Making Way For a Fast Comeback of Red Rice. Justagriculture multidisciplinary e-newsletter Vol.2 Issue-8, 1-6.

Kaur, M., & Iraj, S. (2023). To study a standardisation and development of product of red rice (Oryza sativa) and their organoleptic evaluation. International Journal of Home Science, 9(2): 17-19

M. Kumar, N. Pandey, G.K. Shrivastava, S.C. Mukherjee, And R.R. Saxena, "Grain Yield of Red Rice as Influenced by Integrated Use of Inorganic, Organic and Micro Nutrients," International Journal of Agriculture Sciences, Vol. 10, No. 14, Pp. 6703-6705

Muhidin, J. K., Elkawakib, S., Yunus, M., Kaimuddin, M., Ray, S. G., & Rianda, B. L. (2013). The development of upland red rice under shade trees. World Applied Sciences Journal, 24(1), 23-30.

Niu, Y., Gao, B., Slavin, M., Zhang, X., Yang, F., Bao, J., ... & Yu, L. L. (2013). Phytochemical compositions, and antioxidant and anti-inflammatory properties of twenty-two red rice samples grown in Zhejiang. LWT-Food Science and Technology, 54(2), 521-527.

Rathna Priya, T. S., Eliazer Nelson, A. R. L., Ravichandran, K., & Antony, U. (2019). Nutritional and functional properties of coloured rice varieties of South India: a review. Journal of Ethnic Foods, 6(1), 1-11.

Roy, S., Chakraborty, K., Banerjee, A., Kumar, J., Sar, P., Verma, B. C., & Mandal, N. P. (2023). Evaluation of Rice Germplasm for Tolerance to Multiple Abiotic Stresses Using Multivariate Techniques and Molecular Markers. Indian Journal of Genetics and Plant Breeding, 83(01), 15-23.

Sadimantara, G. R, & Nuraida, W. (2023, September). Evaluation of yield attributing characters and grain yield in some promising red rice lines (Oryza sativa L.) grown in the lowland conditions. In IOP Conference Series: Earth and Environmental Science (Vol. 1241, No. 1, p. 012038). IOP Publishing.

Saxena, A. (2014). Save the Red Rice: A Unique Gift of Nature. International Journal of Current Research in Biosciences and Plant Biology Vol. 1 No. 5, 32-34.

Se, C. H., Chuah, K. A., Mishra, A., Wickneswari, R., & Karupaiah, T. (2016). Evaluating crossbred red rice variants for postprandial glucometabolic responses: a comparison with commercial varieties. Nutrients, 8(5), 308.

Setiowati Y., Hamdani, Husaini M. (2019). Response of Rice Supply in Paser Regency. IOSR Journal of Agriculture and Veterinary Science. Vol. 12, Issue 8 Ser. I, 22-30.

CHAPTER 12

AN ETHNOBOTANICAL SURVEY OF MEDICINAL PLANTS IN GARIABAND DISTRICT, CHHATTISGARH, INDIA

Shubhangi Yadu[1] and Dr. Suneeta Patra[2]

[1,2]Department of Botany, Govt. Nagarjuna P. G. College of Science, Raipur, (C.G.)

Abstract

Plants has enormous potential to treat a wide range of human diseases; these plants are known as medicinal plants. In the Gariaband area of Chhattisgarh, an ethno-botanical study was carried out to record the traditional knowledge and usage pattern of medicinal plants for primary health care by local people, mostly of Baigas and Gonds. The tribal communities still rely on herbal remedies for basic healthcare. During the previous few decades it has seen overexploitation of traditional knowledge and herbal resources, therefore documentation of their knowledge is important so that it can preserved for future generation.

Keywords: Ethnobotany, Ethnomedicine, medicinal plants, Baigas, tribes.

Introduction

Man in the past relied greatly on medicinal plants for curing diseases and illnesses. The quickening pace of life and the demand for fast treatment resulted in the increase of synthetic drugs. However, this artificial substance comes with numerous side effects, issues, and adverse outcomes. This resulted in the resurgence of herbal medicine as a treatment for numerous illnesses. In ancient times, people relied heavily on medicinal plants for healthcare and various needs. According to the World Health Organization (WHO) about 65-80% of the world's population in developing countries depend on plant based drugs for their primary healthcare

(Awoyemi et al., 2012). About 60-80% of the total population in India still rely on plant based crude drugs despite the influx of modern treatments. It has been estimated that folk healers in India use approximately 2500 species of medicinal plants, of which almost 100 species serve as regular sources of medicine (Jain & Patole, 2001; Ved & Goraya, 2008; Sandya & Sandya, 2015; Pandey, 2017). India's diverse climate and environment make it home to some of the richest vegetation in the world. Ethnomedicine indicates the comparison or study of traditional medicinal plants. Chhattisgarh is known as "Herbal state" also it is called as "The land of tribe". Tribal individuals rely on forests for their sustenance, with many still relying on medicinal plants as their main source of healthcare. The indigenous and rural inhabitants of India possess extensive knowledge about the medicinal properties of plants that can be found in their surroundings. The traditional knowledge of using certain medicinal plants is passed down through generations. These plants are of diverse group and are commonly utilized as traditional medicine by rural and tribal communities. Since ancient times, humans have been using plants for medicinal reasons. Due to their substantial significance and commercial value, the sustainable use and conservation of resources are essential.

Study Area

Gariaband district was carved out of Raipur, the capital of Chhattisgarh and has its headquarters at Gariaband town. Dhamtari and Mahasamund is the neighbour district. The district covers an area of 5822.861 sq.km. and is full of natural resources. The district is divided into geographical area of the five talukas Gariaband (726.12 sq.km), Chhura (714.62 sq.km), Mainpur (670.52 sq.km), Devbhog (301.53 sq.km) and Rajim (474.27 sq.km) respectively. The Gariaband, Chhura and Mainpur blocks are area of tribal majority.

Material and Method

The basic data was gathered by the resource person, like herbalist, medicine man (vaidya), elderly people, Baiga, and those with knowledge of medicinal plants and their efficacy from various regions of the Gariaband tribal territory. During the initial study the knowledge about the treatments and alternate methods of treating them was gathered. Multiple visits were done to gather information from the resource persons, who assisted in identifying the plants, their indigenous names, and the extent to which their medical applications were validated. All information regarding the plant parts used, the method of collection and the method of administration and doses were recorded during the investigation.

Results

Following information was gathered and documented from tribal people of study area.

No	Botanical name of plant	Common name	Family	Used part	Used from	Use
1	*Aloe barbadensis* miller	Aloe, Gritakumari	Liliaceae	Stem, leaves	Juice	Dandruff, wounds & Burn, Psoriasis
2	*Andrographis paniculata*	Bhuineem	Acanthaceae	Whole plant	Decoction	Amoebic dysentery, Epilepsy, Leucorrhea
3	*Azadirachta indica*	Neem	Meliaceae	Bark, Leaves, Flower, Seed	Decoction and Oil	Antibacterial, Antifungal, Antimalarial
4	*Bauhinia racemosa Lam*	Raini	Fabaceae	Stem, Bark	Powder	Vomiting, Diarrhea and Blood in stool
5	*Boerhavia diffusa* Linn	Punarnava	Nyctaginaceae	Whole plant	Decoction	Diuretic, Fever, Stimulant, Inflammation Asthma, Anaemia, Headache & Joint problem
6	*Cassia tora* Linn	Charota, Chakavat	Caesalpiniaceae	Leaf, Seed	Decoction	Snake bite, Allergy, Ascaris, Joint pain, Cough
7	*Cocos nucifera* L.	Nariyal	Arecaceae	Fruit	Crushed	Blood purification, Diarrhea and Cold
8	*Dhatura metal* Linn	Dhatura	Solanaceae	Leaf, Seed	Decoction	Asthma, Leprosy, Nacrotic, Respiratory Problems, Toothache, Skin diseases, Fever, Pain relief
9	*Ficus religiosa* L.	Peepal	Moraceae	Stem and Bark	Decoction	Epilipsy, Asthma, Eye pain and Toothache

No	Botanical name of plant	Common name	Family	Used part	Used from	Use
10	*Lagenaria siceraria*	Katutumbi, Kadua	Cucurbitaceae	Fruit	Powder	Joint problem
11	*Lannea coromandelica*	Modga	Anacardiaceae	Stem, Bark	Direct apply	Cuts and Wound
12	*Madhuca longifolia*	Mahua	Sapotaceae	Stem, Bark	Decoction	After delivery, Epilepsy
13	*Meso-sphaerum suaveolens* (L.) Kuntze	Ban tulsi	Lamiaceae	Root, Flower	Decoction	Heart stroke
14	*Moringa oleifera* Lam.	Munga	Moringaceae	Leaves	Boiled	Cold, After delivery and Muscles pain
15	*Ricinus communis*	Arandi	Euphorbiaceae	Seed	Oil	Rheumatism, Purgative, Joint problem
16	*Sida Cardifolia* Linn	Karanti, Bala, Jhanti	Malvaceae	Leaf and Root	Decoction	Diarrhoea, leucorrhoea, Wound, Boils and Joint pain
17	*Solanum nigrum* Linn.	Makoya, Makal	Solanaceae	Whole plant	Decoction and Paste	Diabetes, Laxative, Fever, Eye problem, Joint Problem, Stomach problem
18	*Tinospora cardifolia*	Giloy	Menispermaceae	Leaf, Stem	Direct chewable, Decoction	Fever, Typhoid and Jaundice
19	*Withania sominifera*	Ashwgandha	Solanaceae	Leaf and Root	Decoction	Ulcer, Gynaeco-logical problems, Arthritis, Stimulant, Joint problem
20	*Zingiber officinale* Roscoe.	Adarak	Zingiberaceae	Rhizome	Crushed	Asthma, Cough and Cold, Jaundice, Piles

Discussion

The primary and the most approachable person for tribes for the treatment of their illness is a local healer. They rely on the herbal preparation given by the regional medicine men. The method of herbal preparations includes time and unique techniques of collection of plant parts, single herb or the combination of various herbs, their method of preparations and the dose given. These herbal formulations prepared by Baidas are promising and reliable. From decades the knowledge has been transferred generation after generation and are time tested. These tribes visit nearby hospital or dispensary only when they not don't get relief by herbal preparations or the disease needs serious attention. These indigenous people still primarily receive their healthcare from traditional methods. These valuable traditional knowledge of treating various diseases will be lost if not documented properly.

Conclusion

Since traditional knowledge and Ethnomedical plants are getting lost therefore their documentation is necessary to preserve them. Since traditional knowledge is passed down orally from one generation to the next, the attention for documenting the traditional knowledge is need of hour immediately or it will lose gradually. The treatment of common illnesses like cough, cold, fever, muscle soreness, diarrhoea and for major illnesses and disorders, tribal people still rely on their traditional medicines made from the forest. Chhatisgarh state has a rich biodiversity of medicinal plants a result and has different ethnic people in different region of state. Every ethnic group has its own method of treating different ailments by herbs. Therefore, efforts should be made preserve the knowledge by systematic documentation. This study was an effort to document medicinal plants used by ethnic people for the treatment of various ailments found in Gariaband district, of Chhattisgarh, India.

References

Anonymous, 1985. Ethnobotany of Baigas of Madhya Pradesh, N.B.R.I. Newsletter. 12 : 28-29.

Awoyemi, O.K., Ewa, E.E., Abdulkarim, I. A., & Aduloju, A. R. (2012). Ethnobotanical assessment of herbal plants in south western Nigeria. *Academic Research International*, 2: 50-57.

Brijlal,. & Dube, V.P. (1992). A survey of the plant of ethno medicine of Amarkantak Plateau in CentralIndia. *Agri.Biol. Res.* 8(1):29-37.

Chand, G., Kanungo, V. K., & Naik, M. L.(2021). Ethnomedicinal plants and its Use in Mainpur block of Gariaband district in Chhattisgarh, India. *International Journal of Botany Studies* .6(3): 451-456.

Dwivedi, S.N. (2007). Antimicrobial herbs used among tribal and rural communities. *Ethnobotany*, 19(1&2):111-115.

Dwivedi, S.N., Dwivedi, A., & Dwivedi, S. (2010). Status and utilisation of medicinal plants in Shahdol district, Madhya Pradesh, India, Part 2. *NutraCos* (Milano-Italy), 9: 40-42.

Dwivedi, S.N., Alawa, S. and Mishra, R.P. (2012). Phytochemistry, pharmacological studies and traditional benefits of *Trachyospermum ammi* (Linn.) Sprague. *IJPLS,* 3(5): 1705-1709.

Gupta, A. K., Mishra, S. K. & Khan, A. A. (1999). Ethnobotanical notes on some herbs from Chhattisgarh region of Madhya Pradesh. *Advance Plant Sciences.*12(1): 163-166.

Hingora, I., & Sharma, A. (2017). Traditional uses of medicinal plants of Gariaband District Chhattisgarh. *International Journal of Recent Research in Life Sciences, 4*(1), 9-11.

Jain, A. K., & Patole, S.N. (2001). Less-known medicinal uses of plants among some tribal and rural communities of Pachmarchi forest (M.P.). *Ethnobotany,* 13: 96 -100.

Jain, A. K., (1992). Clans of Sor tribals of Madhya Pradesh and heir role in Conservation. *Ethnobotany.* 4 : 67-69

Jain, B., Jitendra, K., Sheetal, C., & Bhattacharya. A. (2006). Medicinal flora of Madhya Pradesh and Chhattisgarh- A Review. *Indian Journal of Traditional Knowledge,* 5 (2):237-242.

Jain, S.K., & Rao, R.R. (1976). *A Handbook of Field and Herbarium methods.* Today and Tommorow's Printers and Publishers. New Delhi, India. Jain, S.K.(1965). Medicinal plant lore of the tribals of Bastar. *Econ. Bot.* 19: 230-250.

Kokate, C. K., Purohit, A. P., & Gokhale, S. B. (2002). Pharmacognosy (Nirali Publication, Pune), pp. 1-6.

Pandey, G. (2017). Leveraging traditional knowledge for drug discovery. *Journal of Traditional Medicine & Clinical Naturopathy,* 6:2.

Sandya, G. S., & Sandya, K. (2015). Ethno-botanical plants conservation by the tribes of Amarkantak area of District Anuppur, Madhya Pradesh, India. *International Journal of Science and Research,* 4(12):129-1231.

Ved, D. K,. & Goraya, G. S. (2008). Demand and Supply of Medicinal Plants in India, NMPB, New Delhi and FRLHT, Bangalore, India.

CHAPTER 13

A REVIEW ON NUTRITIONAL VALUE OF MILLETS AND IT'S IMPACT ON HUMAN HEALTH

Dr. Madhulika Pandaw[1], Dr. Prakash Kaur Saluja[2]

[1]Department of Botany, Bhakt Mata Karma Government College Samoda, Raipur, C.G.

Email: madhulikapandaw@gmail.com

[2]Department of Botany, Govt. D. B. Girls P.G. (Auto.) College, Raipur, C.G.

Email: prakash.saluja@yahoo.com

Abstract

Millets is an important global crop with significant economic impact in developing countries. Millets belong to the family Poacea which grows well under high temperature and dry condition. It has small seeds and is used as human food and cattle fodder also. Millets is known to be drought resistant crop, they are not prone to pest attack and are also disease resistant. The majority of millets are four to five times more nutritious than the cereals as they have more protein, vitamins, fiber and minerals. Millets are also gluten free. Due to these extraordinary qualities millets have been given the title of "Superfoods". Millets are non-acid forming, it is easily digestible and non-allergic. They contain a large amount of phenolic compounds and fiber. They have medicinal properties such as it is a cure for gastrointestinal ailments like cramps, excessive gas, constipation. Millets have many nutritional properties that are helpful to prevent many health problems like blood pressure, prevention of cancer and cardiovascular diseases. They also help in improving glycemic index and post prandial rise in blood sugar. The millets are boon for curing malnutrition and also in obesity which is major problem in today's modern world. This study focuses on nutritional value of millets and there impact on human health. There nutritional benefits helping in overcoming the world food shortage.

Keywords: Millets, nutritional value, human health, Superfood.

Introduction

Millets belongs to the family Poaceae which can grow well under dry high temperature condition as grasses with small seeds and are considered one of the oldest cultivated crops (Tripathi *et al.* 2021).They are widely grown around the world as cereal crops or grains for human food and fodder. Millets are believed to be originated in Africa. They are important crop in the several tropics of Asia and Africa especially in India, Mali, Nigeria. Millets are often referred to as "coarse cereals" or "cereals for the less affluent."

Millets have deep rooting system and short life cycles and required less water as compared to other cereals crops in terms of the growing period and overall water requirement during growth (Ashwani Kumar *et al.* 2018). Milletes, in contrast to wheat and rice are not only a good sources of energy and major nutrients including protein but are also a good source of micronutrients such as vitamins A, B, D, E, pyridoxine, niacin, zink, iron and antioxidants hence they are known as superfood (Jena *et al.* 2023). They are non acid forming, non allergenic, easily digestible global food with a significant economic impact in developing countries. The crops are grown in a variety of agroecological situation like coasts, planis, hills as well as in varing rainfall and diverse soil conditions. Millets based foods are considered potential prebiotics and probiotics with potential health benefits (Prasanthi and Sireesha 2022).

Millets are short growing season as compared to other major cereals and is known to be drought resistant cropresistance (Devi *et al.* 2011). Globally, there are 13 types of millets, which incuded pearl millet, little millet, finger millet, proso millet, sorghum millet, kodo millet, brown top millet, barnyard millet, foxtail millet, Guinea millet, fonio, teff and job's tears millet. Except for fonio, jobs and teff, the other millets are widely distributed in india (Vetriventhan *et al.* 2020).

Millets are widely used in rural areas. Their grain hold a good quantity of bioactive compounds like polyphenols, tannins, flavonoids, photo-sterol etc. They are having nutritional properties that help in managing health problems like hyperlilidemia, diabetes mellitus (Veena 2003).They have also prevent lowering blood pressure, prevension of cancer, risk of heart diseases and cardiovascular diseases. Alkaline nature of millets help to maintain a healthy pH balance in the body and crucial to prevent illnesses (Sarita and Singh 2016). Many studies is founded that millets contain significant in improving glycine control, post prandial rise in blood glucose concentration and decreasing fasting (Palanisamy and Sree 2020).

Methodology

The review was carried out using the previously reported methodology, with some minor adjustments. The current topic was selected based on literature survey. The objective of the review was evaluate the nutritional value and their impact on human health.

Types of millets

Millets are broadly classified into two types:

1. Major Millets- Sorghum (Jowar) and pearl millet (Bajra)
2. Minor millets- Finger millet (ragi) and small millets such as proso, foxtail, little barnyard and kodo millets

Sorghum

According to molecular studies, It is completely gluten free and can be safely consumed by gluten sensitive individuals. In india and Africa, Sorghum is an ancient cereal grain. For making bread, cookies, pasta etc., Sorgham is best used for the replacement of wheat. Sorghum contains calcium, fibre, iron, protein etc. which help to reducing cholesterol level and other health benefits (Mishra *et al.* 2021).

Pearl millet

It contain good source of fat, insoluble fibre and magnesium. Insoluble fibre helps in reduction of excessive bile system who leads gallstones (Shweta 2015). Magnesium reduces the effects of migraine and improves the respiratory process in asthmatic patients and inhibit the growth of the phytopathogenic fungi (Verma *et al.* 2012).

Finger millet

It contains calcium, iron, fibre, amino acid, proteinwhich is beneficial for infants and pregnant women. It is packed with nutrition which helps to degenerative disease, increase haemoglobin and fight against mal nutrition and inhibit cataract eye lens (Verma *et al.* 2012).

Kodo millets

The presence of phytochemicals, fibre, and antioxidant, helps prevent life style related disorders that are becoming increasingly common today. It is look like rice and used as traditional food. It has easily digestible and help in knee pain, joint pain and women menstruation.

Proso millets

It contain proteins, calcium to strength the bones and dental health. It is also rich in Niacin or vitamins B3 which helps prevent pellagra (Mishra *et al.* 2012), prevent cardiovascular diseases (Verma *et al.* 2012)This crop have grow do not need a lot of water.

Foxtail millets

Due to magnesium content, it helps in preventing diabetes to reducing glucose level in blood and maintains the heart. It causes a consistent glucose release without affecting the bodys metabolism making it diabetic friendly (Reddy 2017).

Banyard Millets

It is rich in phenolic compound and low in sugar that decrease blood sugar spikes after meals. Due to its antioxidant properties and high vitamin content, it is considered a functional food crop (Bhatt, *et al.* 2022).

Nutritional composition of various types of millets

Millets are high in nutrition and dietary fibre. They serve as good sourceof protein, micronutrients and phytochemicals. The millets contain 7-12% protein, 2-5 % fat, 65-75% carbohydrates and 15-20% dietary fibre. The essential amino acid profile of the millet protein is better than various cereals such as maize. Millets contain fewer cross linked prolamins, which may be an additional factor contributing to higher digestibility of the millet proteins. Small millets are good source of phosphorous and iron. Millets contributes to antioxidant activity with phytales, polyphenols, tannins, anthocyanins, phytosterols and pinacosanols present in it having important role in aging and metabolic disease.

Health benefits of millets

Millets have potential health benefits and epidemiological studies have showed that consumption of millets reduces risk of heart disease, protects from diabetes, improves digestive system, lowers the risk of cancer, detoxifies the body, increases immunity in respiratory health, increases energy levels and improves muscular and neural systems and are protective against several degenerative diseases such as metabolic syndrome and parkinsons disease. It is also helpful in kidney and liver diseases. It is also helpful in kidney and liver diseases.The important nutrients present in millets include starch, oligosaccharides, lipids, antioxidants such as phenolic acids, flavonoids, phytosterols which are believed to be responsible for many health benefit.

Challenges for millets cultivation

- Millets are ancient grains that have been cultivated for thousand years. However many people are not aware of their health benefit and nutritional value.
- Millets are considered a low profit crop compared to other crops like wheat and rice and this has discouraged farmers to growing the minutes.
- Millets require proper storage facilities which are not readily available in many areas. This has led to a decline in cultivation (Bhandari *et al.* 2023)
- Processing millets is challenging because of their small seed size and thick seed coat Poor self life of processed grain anf flour
- Any one decortications unit is not suitable for all the millets , due to difference in their grain size, shape and husk content.

Conclusion

This review paper examine the nutritional and health benefits of millets. They exhibits significant potential for various food applications. Due to their drought resistant nature millets can thrive in less fertile lands where other crops may struggles. Millets had numerous health promoting constituents including minerals, fibre, vitamins, pytochemicals, phenolic compound etc. life style related disease and conditions like diabetes and cardiovascular disease can be controlled by consuming millets.

References

Ashwani Kumar; Tomar, Vidisha; Amarjeet Kaur; Vikas Kumar and Gupta, Kritika (2018): Millets: A Solution to Agrarian and Nutritional Challenges. *Agric and Food Secure*, 7(31): 2-15.

Bhandari, Aditya; Kundan, B Jat; Kanse, Aniruddha; Gorde, Prathamesh D; Suri, Shweta and Patil, Anirudh G (2023): Millets As Superfood, A Review. Acta Scientific Nutritional Health, 7 (6): 90-103.

Bhatt, Diksha, Munavirul Fairos and Mazumdar, Ayushi (2022): Millets: Nutritional Composition, Production and Significance: A Review. The Pharma Innovation Journal, 11 (7): 1577- 1582.

Devi, PB; Vijayabharathi R; Sathyabama S; Malleshi NG; Priyadarisini VB (2011): Health Benefits of Finger Millet (Eleusine coracana L.) Polyphenols and Dietary Fibre: A Review. J. Food Sci. Technol

Jena Anurag; Sharma Vishal; Dutta, Usha (2023): Millets As Superfood: Let thy Cereal Be Thy Medicine. Indian Journal of Gastroenterology, 42 (3): 304- 307.

Mishra, Pragya; Prakash, H.G; Suman Devi; Sonkar, Seema; Yadav, Shweta; Singh, H.C. and Singh D. R. (2021): Nutritional Quality of Millets and Their Value Added Products With The Potential Health Benefits: A Review. International Journal of Current Microbiology and Applied Sciences, 10 (10): 163-175.

Palanisamy T, Sree R. (2020): Efficacy of millets (foxtail, kodo, small, barnyard and Pearl millet) varieties on post prandial glycaemic response in patients with type 2 diabetes. Eur J Biomedpharm Sci, 7 : 443-9.

Prasanthi, K and Sireesha G. (2022): Individuals Knowledge, Attitude and Practices on Millets. International Journal of Food and Nutritional Sciences, 11 (3): 21-27.

Reddy O.S.K. (2017): Smart Millet and Human Health, Green Universe Environmental Services Society.

Sarita, Singh, Ekta (2016): Potential of millets: Nutrients Composition and Health Benefits. Journal of Scientific and innovative Research, 5 (2): 46-50.

Tripathi, M. K.; Jodam, R.S., Kumar, A. (2021): Quality Management System in Millet and Sorghum. In Millets and Millet Technology, Springer: Berlin/Heidelberg, Germany 363-379.

Veena B (2003): Nutritional, functional and utilization studies on barnyard millet. M. Science Thesis, University of Agricultural Science, Dharwad (Karnataka), India.

Verma, Veenu and Patel S. (2012): Nutritional Security and Value Added Products From Finger Millets (Ragi). Journal of Applicable Chemistry, 1 (4): 485-489.

Vetriventhan M, Vanica C, Azevedo R, Upadhyaya HD, Nirmalakumari A, Kanepotaka J (2020): Genetic and Genomic Resources and Breeding for Accelerating Improvement of small millets: Current Status and Future Intervention. Nucleus, 63: 217-39.

CHAPTER 14

PARTICIPATORY APPROACHES IN ENVIRONMENTAL PLANNING

Dr. Sangeeta Jha Aditya

Assistant Professor (Economics),
Govt. Arts and Commerce Girls' College, Devendra Nagar, raipur Chhattisgarh
sangaditya222@gmail.com

Abstract

Participatory approaches in environmental planning have become increasingly important as traditional top-down methods fail to adequately address the complexities of environmental issues and the needs of affected communities. This chapter explores the evolution, theoretical foundations, and practical applications of participatory approaches in environmental planning. Beginning with a historical overview, it traces the shift from expert-driven decision-making to inclusive, community-centered processes. The chapter delves into the various methods and tools employed to facilitate participation, including public consultations, focus groups, participatory mapping, and digital platforms. Challenges such as power imbalances, representation issues, and resource constraints are examined, alongside strategies to overcome these obstacles. Case studies highlight the impact of participatory planning on environmental outcomes, demonstrating its potential to enhance sustainability, equity, and resilience. The chapter concludes by discussing emerging trends and innovations in participatory environmental planning, emphasizing the need for continued evolution and adaptation in this field.

Keywords: Environmental Planning, Public Participation, Stakeholder Engagement, Sustainable Development, Environmental Governance.

Introduction

Environmental planning plays a crucial role in managing natural resources, conserving biodiversity, and mitigating environmental impacts in a rapidly changing world. Historically, environmental planning was often a top-down process, dominated by experts and policymakers. However, this approach frequently overlooked the insights, needs, and values of the communities directly affected by environmental decisions. In response to these shortcomings, participatory approaches have gained prominence, emphasizing the importance of including diverse stakeholders in the planning process.

Participatory approaches in environmental planning involve engaging local communities, indigenous peoples, non-governmental organizations (NGOs), and other relevant stakeholders in decision-making. These approaches aim to democratize environmental governance, ensuring that those who are most affected by environmental policies have a voice in shaping them. This chapter explores the evolution, methods, challenges, and impacts of participatory approaches in environmental planning, highlighting the transformative potential of inclusive decision-making.

Historical Context and Evolution of Participatory Approaches

Early Environmental Planning Practices:

The roots of environmental planning can be traced back to early conservation movements, where the focus was primarily on protecting natural resources from over-exploitation. Initially, environmental planning was characterized by a top-down approach, where decisions were made by a few experts without significant input from the broader public. This model often led to conflicts, as it failed to account for the social, cultural, and economic realities of the affected communities.

Emergence of Participatory Models:

The limitations of the top-down approach became increasingly evident during the latter half of the 20th century, particularly as environmental degradation and social inequalities grew more pronounced. In response, participatory models began to emerge, emphasizing the need for more inclusive and democratic planning processes. The 1987 Brundtland Report, "Our Common Future," played a pivotal role in highlighting the importance of public participation in sustainable development. It argued that sustainable development could only be achieved if local communities were actively involved in the decision-making processes that affected their lives.

Case Studies from Different Regions:

The evolution of participatory environmental planning can be seen in various regional case studies. For instance, in Latin America, participatory planning emerged as a response to authoritarian regimes that excluded marginalized communities from decision-making. In Africa, participatory approaches have been used to integrate indigenous knowledge systems into natural resource management, leading to more sustainable and culturally appropriate outcomes. In Europe, the Aarhus Convention (1998) institutionalized the public's right to participate in environmental decision-making, setting a global precedent for participatory environmental governance.

Theoretical Framework of Participation

Defining Participation:

Participation in environmental planning is a complex and multi-dimensional concept, encompassing various forms of engagement ranging from passive information sharing to active decision-making. At its core, participation seeks to redistribute power, enabling communities and stakeholders to influence outcomes that directly affect them. This shift towards participatory approaches reflects a broader trend towards democratizing governance and enhancing social equity in environmental decision-making.

Arnstein's Ladder of Citizen Participation:

A foundational framework for understanding participation is Sherry Arnstein's "Ladder of Citizen Participation," which categorizes different levels of citizen involvement in decision-making processes. The ladder ranges from "manipulation" and "therapy," where participation is merely a formality, to "citizen control," where stakeholders have full authority over decisions. This framework highlights the importance of moving beyond tokenistic forms of participation towards genuine empowerment of communities.

The Spectrum of Participation in Environmental Planning:

In environmental planning, the spectrum of participation can be seen in various forms, from consultation and collaboration to co-production and delegation of power. Consultation involves gathering input from stakeholders, while collaboration and co-production involve stakeholders working alongside experts and policymakers to develop solutions. Delegation of power, on the other hand, allows communities to take the lead in decision-making, with governments and experts providing support.

Empowerment and Inclusion in Decision-Making:

True participation goes beyond mere consultation; it involves empowering marginalized communities to take an active role in decision-making processes. This requires creating spaces where diverse voices can be heard and ensuring that power dynamics do not silence or marginalize certain groups. Empowerment also involves building the capacity of communities to engage meaningfully in environmental planning, through education, training, and access to information.

Methods and Tools for Participatory Environmental Planning

Public Consultations

Public consultations are one of the most common methods of involving stakeholders in environmental planning. These consultations can take various forms, including public meetings, hearings, and surveys. While public consultations provide a platform for stakeholders to express their views, their effectiveness depends on how well the input is integrated into the final decisions. Ensuring that consultations are inclusive, accessible, and transparent is crucial for their success.

Focus Groups and Workshops

Focus groups and workshops provide more interactive and in-depth opportunities for stakeholder engagement. These methods allow for smaller, more focused discussions, where participants can explore issues in greater detail and collaboratively develop solutions. Workshops, in particular, can be used to bring together diverse stakeholders, including community members, experts, and policymakers, to co-create environmental plans.

Participatory Mapping and GIS

Participatory mapping and Geographic Information Systems (GIS) are powerful tools for involving communities in environmental planning. These methods allow stakeholders to visually represent their knowledge and perspectives on environmental issues, such as land use, natural resource management, and conservation. Participatory mapping can help bridge the gap between local knowledge and scientific data, leading to more comprehensive and context-specific planning outcomes.

Digital and Online Platforms

In the digital age, online platforms and tools have become increasingly important for facilitating participation in environmental planning. Digital platforms can enhance transparency, broaden participation, and provide real-time feedback. Examples include online forums, social media platforms, and specialized environmental planning software. However, digital participation also raises challenges related to accessibility, digital literacy, and data security.

Scenario Planning

Scenario planning is a method that involves creating and analyzing multiple potential future scenarios to inform decision-making. This approach allows stakeholders to explore the implications of different decisions under various conditions, helping them to identify the most resilient and sustainable strategies. Scenario planning is particularly useful in environmental planning, where uncertainty and complexity are often significant challenges.

Challenges and Opportunities in Participatory Planning

Common Challenges: Power Dynamics, Representation, and Resource Constraints

Despite the potential benefits of participatory approaches, there are several challenges that can undermine their effectiveness. Power dynamics often play a significant role, with more powerful stakeholders dominating the process and marginalizing less influential groups. Representation is another critical issue, as ensuring that all relevant stakeholders, particularly those from marginalized communities, are included in the process can be difficult. Additionally, participatory processes can be resource-intensive, requiring significant time, funding, and expertise.

Addressing Challenges: Strategies for Effective Participation

To address these challenges, it is essential to adopt strategies that promote inclusivity, equity, and transparency in participatory planning. This includes actively seeking out and engaging marginalized groups, providing capacity-building support to enable meaningful participation, and establishing clear mechanisms for integrating stakeholder input into decision-making. Ensuring that participatory processes are well-resourced and adequately supported is also critical to their success.

Opportunities: Innovation in Participatory Tools and Processes

Advancements in technology and methodology have opened new opportunities for enhancing participation in environmental planning. Innovations such as participatory GIS, digital platforms, and mobile apps have made it easier to involve a broader range of stakeholders and to gather and analyze data in real-time. These tools can help to overcome some of the traditional barriers to participation, such as geographic distance and resource constraints, making the planning process more inclusive and responsive.

Case Study: Participatory Planning in Urban Green Spaces

Urban green spaces are essential for the ecological health and social well-being of cities. They offer numerous benefits, such as improving air quality, providing recreational areas, enhancing biodiversity, and contributing to the overall aesthetic of urban environments. However, the planning and management of these spaces can often be contentious, particularly when the needs of local communities are overlooked. Participatory planning, which actively involves residents and stakeholders in the decision-making process, has emerged as a key strategy to ensure that urban green spaces are designed and managed in ways that meet the diverse needs of the communities they serve.

This case study examines participatory planning in urban green spaces through the lens of a specific project: the transformation of the Bywater neighborhood's green spaces in New Orleans, Louisiana. This example highlights the challenges and successes of involving the community in the planning process and the long-term benefits of such an approach.

Background: Bywater Neighborhood, New Orleans

The Bywater neighborhood is a historically significant area of New Orleans, known for its vibrant culture, historic architecture, and strong community ties. However, like many urban areas, By water has faced challenges related to gentrification, environmental degradation, and the loss of public green spaces. In the early 2000s, the community identified a need to revitalize its green spaces to improve the quality of life for residents and to preserve the neighborhood's unique character.

The Participatory Planning Process

Recognizing the importance of community input, local government officials, urban planners, and non-profit organizations initiated a participatory planning process to redesign and enhance Bywater's green spaces. The process involved several key steps:

1. **Community Engagement:** The first phase of the project involved extensive community outreach. This included public meetings, surveys, and focus groups to gather input from a wide range of stakeholders, including residents, business owners, and environmental organizations. Efforts were made to ensure that all segments of the community, particularly those who might be marginalized or less likely to participate, had a voice in the planning process.
2. **Needs Assessment and Visioning:** Based on the feedback collected, the planning team conducted a needs assessment to identify the community's priorities. This led to the development of a shared vision for the neighborhood's green spaces, which focused on creating areas that would serve as communal gathering spots, support local biodiversity, and provide recreational opportunities for all age groups.
3. **Collaborative Design:** With the community's vision in mind, landscape architects and urban planners worked alongside residents to create detailed designs for the green spaces. Participatory tools such as interactive mapping sessions, design charrettes, and model-building workshops were used to help residents visualize potential changes and provide feedback on different design options.
4. **Implementation:** The community's involvement continued during the implementation phase. Local residents volunteered to assist with planting trees, installing benches, and creating community gardens. This hands-on participation fostered a sense of ownership and pride in the new green spaces.
5. **Ongoing Management:** After the green spaces were completed, a community stewardship program was established to ensure their ongoing maintenance and management. This program, led by a coalition of local residents and non-profit organizations, continues to organize volunteer activities and events in the green spaces, ensuring they remain vibrant and well-cared for.

Outcomes and Impact

The participatory planning process in Bywater resulted in several positive outcomes:

- **Enhanced Green Spaces:** The revitalized green spaces now include community gardens, play areas for children, walking paths, and areas for cultural events and performances. These spaces have become focal points for community life, providing a much-needed respite from the urban environment.

- **Strengthened Community Ties:** The process of working together to design and implement the green spaces helped to strengthen the sense of community in Bywater. Residents who might not have interacted before the project found common ground in their shared vision for the neighborhood.
- **Increased Environmental Awareness:** The project also raised awareness about the importance of urban green spaces for environmental sustainability. Educational programs about native plants, urban wildlife, and sustainable gardening practices were incorporated into the design of the green spaces.
- **Economic Benefits:** The improvements in Bywater's green spaces have contributed to an increase in property values and have attracted new businesses to the area, contributing to the neighborhood's economic revitalization.

Challenges and Lessons Learned

While the participatory planning process in Bywater was largely successful, it was not without challenges:

- **Balancing Diverse Interests**: One of the main challenges was balancing the diverse interests of different community groups. For example, some residents prioritized recreational facilities, while others wanted to focus on preserving green areas for wildlife. Negotiating these different priorities required ongoing dialogue and compromise.
- **Resource Limitations**: Funding and resources were limited, which meant that not all community desires could be realized. The planning team had to be creative in finding low-cost solutions and securing additional funding through grants and donations.
- **Gentrification Concerns**: The improvements to Bywater's green spaces also raised concerns about gentrification and the potential displacement of long-time residents. While the project aimed to benefit the existing community, it also contributed to an increase in property values, which could lead to affordability issues.

Despite these challenges, the Bywater project offers valuable lessons for other communities seeking to engage in participatory planning for urban green spaces:

1. **Inclusive Engagement is Key**: Ensuring that all voices are heard, particularly those of marginalized groups, is crucial for the success of participatory planning. This requires proactive outreach and the use of diverse engagement methods.
2. **Flexibility and Adaptability**: The planning process must be flexible enough to adapt to changing circumstances and the evolving needs of the community. Continuous feedback loops and iterative design processes can help accommodate these changes.

3. **Long-Term Commitment**: Successful participatory planning extends beyond the initial design and implementation phases. Long-term management and community involvement are essential to maintaining and enhancing urban green spaces.

The Bywater case study illustrates the potential of participatory planning to create urban green spaces that are responsive to the needs of local communities, environmentally sustainable, and socially inclusive. By involving residents in every stage of the planning process, the project not only enhanced the physical environment but also strengthened community ties and fostered a sense of ownership and pride in the neighborhood's green spaces. While challenges remain, the Bywater experience provides a model for other urban areas looking to engage in participatory planning for the creation and management of green spaces.

References

Arnstein, Sherry R. (1969). "A Ladder of Citizen Participation." *Journal of the American Institute of Planners*, 35(4), 216-224.

United Nations (1987). "Our Common Future: Report of the World Commission on Environment and Development."

Reed, Mark S. (2008). "Stakeholder participation for environmental management: A literature review." Biological Conservation, 141(10), 2417-2431.

Fiorino, Daniel J. (1990). "Citizen Participation and Environmental Risk: A Survey of Institutional Mechanisms." Science, Technology, & Human Values, 15(2), 226-243.

Berkes, Fikret. (2009). "Evolution of co-management: Role of knowledge generation, bridging organizations and social learning." Journal of Environmental Management, 90(5), 1692-1702.

International Association for Public Participation (IAP2). (2018). "IAP2 Spectrum of Public Participation."

Fung, Archon. (2006). "Varieties of Participation in Complex Governance." Public Administration Review, 66, 66-75.

Conrad, Esther, & Hilchey, Krista G. (2011). "A review of citizen science and community-based environmental monitoring: issues and opportunities." Environmental Monitoring and Assessment, 176(1-4), 273-291.

Agyeman, Julian, & Evans, Bob. (2003). "Toward Just Sustainability in Urban Communities: Building Equity Rights with Sustainable Solutions." The ANNALS of the American Academy of Political and Social Science, 590(1), 35-53.

Tippett, Joanne, Handley, John F., & Ravetz, Joe. (2007). "Meeting the challenges of sustainable development—A conceptual appraisal of a new methodology for participatory ecological planning." Progress in Planning, 67(1), 9-98.

CHAPTER 15

ENHANCING ENVIRONMENTAL EDUCATION AND PUBLIC AWARENESS: STRATEGIES FOR SUSTAINABLE DEVELOPMENT

Bhaskar Chandra[1], Om Kumari Sahu[1]
Dr. Kavita Sharma[2], Dr. B.M. Lall[1]

[1]Department of Botany Govt. D.B. Girls College Raipur (C.G.)
[2]Department of Botany Govt. Arts & Commerce Girls College (C.G)
Email: bhaskarchandra893@gmail.com

Abstract

The urgency of addressing environmental issues such as climate change, pollution, and resource depletion is increasingly evident amid global concerns over rising temperatures and environmental degradation. Modern technologies and changing human lifestyles amplify these challenges, particularly in developing countries facing escalating pollution levels. Environmental education (EE) emerges as a pivotal strategy to mitigate these impacts by fostering awareness, knowledge, and responsible behaviors from an early age. Rooted in various epistemologies, EE aims to integrate understanding of natural and social systems, empowering individuals to engage in sustainable practices and advocate for environmental stewardship. This paper examines the role of EE in promoting sustainable development through enhanced environmental awareness and education. It explores the concepts of awareness, knowledge, participation, and behavioral change as fundamental pillars of EE, emphasizing their collective impact on fostering environmental stewardship and informed citizenship. Additionally, the paper discusses the complementary role of public awareness campaigns in shaping attitudes and behaviors towards sustainability, highlighting their influence on policy-making and global cooperation. Challenges in implementing EE, such as professionalism and addressing climate change complexities, underscore the need for robust educational frameworks and innovative methodologies. Case studies like the Eco-Schools program demonstrate effective EE practices in cultivating pro-environmental attitudes among students and local communities. Moreover, recent global initiatives like the Berlin Declaration underscore the imperative to integrate education for sustainable development (ESD) into educational systems worldwide, emphasizing collaborative action towards achieving sustainable development goals.

Keywords: - Environmental Education, Public Awareness, Sustainable Development, Natural Resource Preservation, Climate change.

Introduction

The escalating consciousness of environmental issues underscores the pervasive impacts of climate change, evidenced by rising global temperatures, mounting pollution, and the depletion of non-renewable natural resources. (Douglas, 2006). Modern technologies and shifting lifestyle patterns among humans significantly threaten the environment. Pollution levels in developing countries are escalating daily. Consequently, there is an increasing need for environmental education to enhance environmental quality and foster sustainable development. Education can equip students with knowledge and awareness, leading to behavioral changes that support the maintenance of a sustainable environment. (Hassan et al., 2011). The consideration of environmental education and social change is rooted in a philosophy or theory of knowledge that elucidates the functioning and interaction of the natural and social worlds. Different theories of knowledge, called epistemologies, provide varying perspectives on what can be known (ontologies) and the methods for acquiring information and knowledge (methodologies). Realism is an epistemology that emerged as a counter to idealism, which posits that reality exists only in our perceptions. Idealists believe that reality is entirely a mental construct, whereas realists maintain that there are independent structures and processes in the world that generate phenomena regardless of our ability to perceive or access them. (Sayer, 1985; (Gregory, 1986).

Sustainable development involves utilizing the environment for growth while simultaneously protecting and conserving it, ensuring that it can meet human needs both now and in the future. (Nursadiah et al., 2018). Environmental issues can only be addressed by increasing awareness about them. Hence, one of the most effective ways to promote environmental protection and enhance environmental awareness is by bolstering environmental education in primary and junior high schools. (Ergin, 2019) Education serves as the catalyst for social engineering within society, leading to transitions and changes such as shifts in values, modes of thinking, material culture, and cultural norms. This includes fostering knowledge and attitudes toward the environment, influencing material culture patterns, ideas, family relations, political culture, and administrative structures at various levels. Through education, individuals become aware of the behaviors deemed acceptable within their culture and learn to modify their behavior accordingly. This process equips individuals with adaptive capacities, enabling them to adjust to societal changes. (Ezimah, 2021).

Definition

Environmental education refers to the organized efforts to teach individuals and communities about the natural environment and how to manage their behavior and ecosystems sustainably. It aims to develop a population that is knowledgeable about the environment, aware of environmental challenges, and motivated to engage in environmental problem-solving.

Concepts

1. Awareness and Sensitivity

- Fostering an awareness of the environment and its associated challenges.
- Cultivating sensitivity towards the environment and recognizing the impact of human actions.

2. Knowledge and Understanding

- Providing information and understanding about the natural environment and its processes.
- Educating on the interdependence between humans and the environment.

3. Participation

- Encouraging active participation in activities that lead to the resolution of environmental challenges.
- Fostering cultivating an ethic of accountability and care for the natural world.

4. Behavioral Change

- Aiming to influence behavior to support sustainable and eco-friendly practices.
- Promoting lifestyle changes that reduce environmental impact and promote sustainability.

Importance of Environmental Education

- **Empowerment**: Empowers individuals to take informed actions and make responsible decisions regarding environmental issues.
- **Sustainable Development**: Supports the goals of sustainable development by promoting an understanding of the need to balance environmental, social, and economic factors.
- **Informed Citizenship**: Cultivates informed and engaged citizens who can participate effectively in environmental decision-making processes.

Long-term Impact: Establishes a foundation for long-term environmental stewardship and conservation efforts.

The Importance of Public Knowledge in Preserving the Environment

Public awareness plays a crucial role in environmental conservation by influencing individual and collective actions towards sustainable practices and the protection of natural resources.

Education and Understanding

Public awareness campaigns educate people about environmental issues, such as climate change, biodiversity loss, and pollution.

Increased understanding fosters a sense of responsibility and empowers individuals to make informed decisions.

Behavioral Change

Awareness campaigns encourage individuals to adopt eco-friendly behaviors, such as reducing waste, conserving energy and water, and using sustainable transportation. Public awareness helps in shaping consumer choices towards environmentally friendly products and services.

Advocacy and Engagement

Awareness fosters advocacy for environmental policies and regulations that promote conservation efforts. It encourages public engagement in community-based initiatives, conservation projects, and environmental activism.

Policy Influence

Informed and engaged citizens can influence policymakers to prioritize environmental conservation and sustainability in legislation and governance. Public pressure and support for environmental causes can lead to the implementation of effective environmental policies.

Long-term Impact

Sustainable changes in public behavior contribute to long-term environmental conservation and preservation of natural habitats. Public awareness helps in building a culture of environmental stewardship and collective responsibility for future generations.

Global Perspective

Awareness campaigns can highlight global environmental challenges and promote international cooperation and agreements for addressing issues like climate change and biodiversity conservation.

Importance of Environmental Education

Environmental education is crucial for raising awareness about the importance of conservation, sustainability, and environmental protection. It empowers individuals to make informed decisions and take action to mitigate environmental issues, such as climate change, pollution, and biodiversity loss.

Environmental education has numerous benefits, including:

1. Promoting sustainable behaviors and lifestyles
2. Encouraging responsible resource management
3. Fostering a sense of environmental stewardship and responsibility
4. Supporting the development of eco-friendly technologies and practices
5. Enhancing ecological literacy and critical thinking skills
6. Encouraging community involvement and participation in environmental decision-making
7. Preparing future generations to address environmental challenges

Strategies for Enhancing Public Awareness

Enhancing public awareness of environmental education is crucial for inspiring action and promoting sustainability. Here are some strategies to consider:

1. Media Campaigns: Leverage social media, television, radio, and print media to reach a wider audience.
2. Community Events: Organize events like festivals, fairs, and workshops to engage the public and make environmental education accessible.
3. Collaborate with Influencers: Partner with social media influencers, thought leaders, and celebrities to amplify environmental messages.
4. Education Centers and Museums: Establish interactive centers and museums to provide hands-on environmental learning experiences.
5. Public Art and Installations: Use art to convey environmental messages and create public installations that raise awareness.
6. School Programs: Integrate environmental education into school curricula and activities to reach students and teachers.
7. Community Outreach: Engage with local communities, businesses, and organizations to promote environmental education and involvement.
8. Social Media Challenges: Launch viral social media challenges that encourage individuals to take environmental actions.
9. Public Lectures and Workshops: Host expert lectures and workshops to educate the public on environmental topics.

10. Celebrate Environmental Days: Organize events and campaigns around international environmental days, such as Earth Day and World Environment Day.
11. Develop Engaging Content: Create informative and engaging content, such as videos, podcasts, and blogs, to share environmental knowledge.
12. Partner with Businesses: Collaborate with eco-friendly businesses to promote environmental education and practices.

Challenges in Promoting Environmental Education

Some challenges in promoting environmental education include.

Deficiency in professionalism

Numerous challenges arise during the implementation of EE. A prevalent issue is the lack of professionalism, which can be either subtle or overt. It is suggested that EE programs need to be effectively designed and executed, with specialists in pedagogy at the helm, and that these programs should be enhanced through evidence-based research (Marcinkowski, 2009). Despite significant progress in EE development since the 1990s, particularly in professional programming and development, the advancement is still not at its peak. EE has long faced criticism for being advocacy-driven and for delivering messages that are emotionally charged, pessimistic, and sometimes inaccurate. (Marcinkowski, 2009). EE has been criticized for its shortcomings in economics and science and for being contentious due to its failure to adequately equip learners with the skills to understand trade-offs in addressing complex environmental issues. Critics have raised concerns about the credibility of underqualified EE providers, the use of superficial and inaccurate materials, the program's advocacy practices, and the limited scope of its offerings. This perspective questions the overall effectiveness and reliability of EE programs. (Salmon, 2000).

The issue of climate change

A third challenge in implementing EE is the issue of climate change. The national evaluation of professional needs in environmental literacy identified climate change as a critical content area. However, according to Marcinkowski (2009), teaching climate change concepts presents unique pedagogical and methodological challenges for EE learners. Climate change stands as the foremost environmental crisis, demanding societal attention and education due to the escalating loss of biodiversity, ozone depletion, population growth, pollution, and resource depletion. Meanwhile, EE aims to influence behavior and lifestyles through training and educational programs to address these environmental issues. The challenge lies in how effectively EE can be applied to climate change adaptation and mitigation. For example, while East African countries such as Kenya, Uganda, and Tanzania have made strides in promoting EE in the region, the progress has been undermined by the severe weather conditions intensified by climate change (Johnson-Pynn & Johnson, 2005).

The Impact of EE on Environmental Sustainability:-

According to Stanford University experts (2013), they extensively documented the outcomes of EE concerning environmental sustainability in a comprehensive review. In addition to raising environmental awareness, EE has yielded several positive impacts. These include enhancing academic performance, improving critical thinking and reasoning skills, fostering life skills and personal growth. Moreover, learners have demonstrated increased engagement in civic duties and exhibited positive environmental behaviors. In their review study, Stanford University experts presented evidence suggesting that EE learning offers a transformative opportunity to advance environmental sustainability. (Ardoin et al., 2018). The study indicated that EE learning fosters confidence, skills, behavior, motivation, and improved academic performance among learners, encouraging sustainable environmental practices. (Stern et al., 2011).

Case studies and best practices of environmental education (EE):-

'Eco-Schools' is a global school program involving numerous schools worldwide in structured activities aimed at cultivating pro-environmental attitudes and behaviors among students. This inclusive program extends its activities beyond pupils and school communities to involve local communities as well. According to the Eco-Schools webpage, the program comprises three main components: The Seven Steps Framework, Eco-Schools Themes, and Assessment for the Green Flag. Implementation of the program requires support from school leaders and the board, with active participation from staff. (Knippels et al., 2021)

To illustrate specific successful practices within this extensive program, we examine the case of Llanmiloe Community Primary School in Wales. In this case study, the program is funded by the Welsh government and coordinated by the National Office of "Keep Wales Tidy". (Knippels et al., 2021)

Aim of School:-The goal is for schools to initiate a significant journey towards enhancing the environment within both the school and the local community, thereby creating a lasting positive influence on the lives of young people, their families, school staff, and local authorities. (https://www.ecoschools.global/how-does-it-work)

Uses:-The Eco-Schools program utilizes action-based learning methods, which encompass a variety of activities. These include participating in global action campaigns such as the "One Planet picnic," engaging in regular outdoor learning sessions like "Outdoor Classroom Day" and the "Surf to Turf - Fundraising Cycle Ride," as well as participating in eco-friendly online activities such as learning about fruits and vegetables from different parts of the world. (Knippels et al., 2021)

Future Directions of EE

Over 80 ministers and vice ministers, along with 2,800 education and environment stakeholders, committed to advancing education for sustainable development (ESD) by adopting the Berlin Declaration at a virtual World Conference organized by UNESCO in cooperation with the Federal Ministry of Education and Research of Germany and the German Commission for UNESCO. The conference, attended online by over 10,000 viewers, emphasized integrating ESD into education systems worldwide by 2025.

UNESCO highlighted gaps in current educational plans, noting that more than half of them do not address climate change and only 19% mention biodiversity. The Berlin Declaration outlines policies to transform learning through teaching, professional training, and civic engagement, emphasizing cognitive, social, and emotional skills, collaboration, problem-solving, and resilience-building. German Chancellor Angela Merkel stressed the importance of making sustainable development education accessible to all, linking its success to achieving the Sustainable Development Goals (SDGs). Discussions also underscored the role of education in combating climate change, echoing commitments made in the Paris Agreement. The voices of young people advocating for transformative education were prominent throughout the conference, emphasizing the need for education to equip them to protect and sustain the planet. The adoption of the Berlin Declaration sets the stage for implementing the ESD for 2030 Roadmap, urging UNESCO Member States to create networks for collaborative action toward achieving ambitious educational goals. Opportunities in 2021, including key global conferences on biodiversity and climate change, will further drive global efforts to integrate ESD into educational frameworks worldwide.

Conclusion

In conclusion, enhancing environmental education (EE) and fostering public awareness are crucial strategies for achieving sustainable development. By integrating comprehensive EE into primary and junior high school curricula, societies can cultivate a culture of environmental consciousness from an early age. This approach not only empowers individuals with knowledge and awareness but also promotes environmentally responsible behaviors essential for the sustainable use and preservation of natural resources. Public awareness initiatives, including community engagements, media campaigns, and policy interventions, further amplify these efforts by educating and mobilizing individuals and communities to adopt sustainable practices. Together, strengthened EE and heightened public awareness can contribute significantly to building a sustainable future for current and future generations.

References

Ardoin, N. M., Bowers, A. W., Roth, N. W., & Holthuis, N. (2018). Environmental education and K-12 student outcomes: A review and analysis of research. *The Journal of Environmental Education*, *49*(1), 1-17.

Douglas, J. (2006). Building adaptation. Routledge.

Ergin, D. Y. (2019). Environmental Awareness of Teacher Candidates. World Journal of Education, 9(1), 152-161.

Ezimah, M. O. A. (2021). The imperatives of environmental education for sustainable development. J. Adv. Educ. Philos, 5, 98-102.

Gregory, D. (1986), 'Realism', in R. J. Johnston, D. Gregory & D. M. Smith (eds), Th~ Dictionary of Human Geography, 2nd edn, Basil Blackwell, Oxford.

Hassan, A., Rahman, N. A., & Abdullah, S. I. S. S. (2011). The level of environmental knowledge, awareness, attitudes and practices among UKM students. University Kebangsaan, Malaysia, 13, 5-8.

https://www.unesco.org/en/articles/unesco-declares-environmental-education-must-be-core-curriculum-component-2025

Johnson-Pynn, J. S., & Johnson, L. R. (2005). Successes and challenges in East African conservation education. The Journal of Environmental Education, 36(2), 25-39.

Knippels, M. C., Romero Ariza, M., Boeve-de Pauw, J., Gericke, N., Paraskeva-Hadjichambis, D., Telesiene, A., ... & Hadjichambis, A. C. (2021). Report on good examples and best educational practices leading to pro-environmental attitudes, behaviour and values.

Marcinkowski, T. J. (2009). Contemporary challenges and opportunities in environmental education: Where are we headed and what deserves our attention?. The journal of environmental education, 41(1), 34-54.

Nursadiah, Suyana, I., & Ramalis, T. R. (2018). Profil sustainability awareness siswa melalui integrasi ESD dalam pembelajaran berbasis masalah pada topik energi di SMP. Prosiding Seminar Nasional Fisika (SINAFI), 207–212.

Salmon, J. (2000). Are we building environmental literacy?. The Journal of Environmental Education, 31(4), 4-10.

Sayer, A. (1985). Realism and Geography'in Johnston, RJ (ed.), The Future of Geography.

Stern, M. J., Powell, R. B., & Ardoin, N. M. (2010). Evaluating a constructivist and culturally responsive approach to environmental education for diverse audiences. The Journal of Environmental Education, 42(2), 109-122.

CHAPTER 16

HUMAN BEHAVIOR AND SUSTAINABLE TECHNOLOGY ADOPTION

Dr. Mini Alex[1] & Ms. Christin Mary Alex[1]

[1]*Govt D.B.P.G. Girls College, Raipur, C.G.*

Abstract

In an era filled with degradation of the environment and climate change, sustainability has emerged as a global priority. In this endeavour, technology is essential since it provides creative ways to reduce our environmental impact. However, human behavior—the decisions we make, the habits we establish, and the values we uphold—affects these technologies just as much as their functioning and design. This article explores the intricate relationship between human behavior and the adoption of sustainable technologies.

Keywords: Sustainability, Technology, Human behavior.

Introduction

Sustainability aims to enable human habitation on Earth while preserving its natural resources. The ultimate objective is to build a better future for the earth and its inhabitants. Innovations that support social and economic advancement while also taking into account the environment are known as sustainable technology. With sustainable results, these technologies seek to dramatically reduce hazards to the environment and ecology. It involves mitigating environmental damage, increasing energy and resource efficiency, and replacing non-renewable materials with renewable ones.

Sustainable technology involves a wide array of innovations and efforts aimed at minimizing environmental impact while boosting efficiency and productivity. It represents a significant shift in the direction of technology development and usage, emphasizing waste reduction, renewable energy sources, and eco-friendly methods.

This movement goes beyond simply reducing the negative effects of current technologies; it is also about rethinking and redesigning future advancements with sustainability as a core principle.

The importance of sustainable technology in today's world goes beyond just protecting the environment. It is about finding a balance where technological advancement does not harm the planet's well-being. This idea is increasingly becoming a key element in corporate strategies, government policies, and consumer choices. As we move closer to the digital age, incorporating sustainability into technology advancements is becoming more and more important. It affects how we work, live, and interact with our surroundings. (Brandenburg G, 2023)

Consumer decisions are influenced by human behaviour, which in turn reflects and strengthens the underlying attitudes and behaviours about sustainability. The growing emphasis on sustainability by consumers can lead to wider market shifts that impact the creation and accessibility of sustainable goods and innovations.

There is a cyclical relationship between consumer decisions and the development of sustainable technologies. Businesses are motivated to invest in eco-friendly technologies as consumers want more sustainable products and become more aware of environmental issues. As a result of these investments, the market offers a greater variety of sustainable solutions, further enabling customers to make eco-friendly decisions.

Therefore, while technological advancements are essential in the fight against environmental degradation, they alone are not enough. The way something is supposed to function in theory often differs from how it operates in reality, this is where human behavior comes in. It means that the success of these innovations depends on the willingness of individuals to embrace them, alter their habits, and integrate them into their daily lives.

Navigating Obstacles & Facilitating Change

Human behaviour is the root cause of many environmental problems, including littering, water scarcity, and air pollution (Gardner & Stern, 2002). We must change the behaviours that lead to these problems if we are to improve the quality of the environment. For example, sustainable technologies require not only technical innovations for energy efficiency but at the same time, individuals need to accept and understand them, buy them, and use them in proper ways. Adoption of sustainable technology in successful cases frequently involves behavioural methods. For example, government laws, public awareness efforts, and technology breakthroughs that have made LEDs more accessible and affordable have propelled the global shift to energy-efficient LED lighting. Information on cost savings and gradual phase-out helped overcome initial resistance to the switch from incandescent bulbs.

However, many obstacles stand in the way of the broad adoption of sustainable technologies, with human behavioural barriers being one such obstacle. Adoption of sustainable technologies is significantly hampered by behavioural and cognitive hurdles, which are ingrained in our daily routines and psychological inclinations. One of the main problems is resistance to change; People often stick to familiar routines due to comfort and predictability. Cognitive biases, such as the status quo bias, which causes people to favour the status quo even in the face of superior options, can increase this reluctance. Resistance might also result from a lack of knowledge about the advantages of sustainable technologies. Either they don't realise the beneficial effects these technologies can have on people's lives and the environment, or they don't realise the long-term benefits. Cognitive biases, such as present bias, prioritize immediate gratification over long-term benefits, making the initial investment in sustainable technologies seem less appealing despite future savings and environmental advantages.

Targeted strategies, such as incentives, education, and human-centered design, are needed to remove these obstacles. According to the Theory of Planned Behaviour (Ajzen, 1991), people make decisions based on attitudes, perceived control of behaviour, and social norms, and they select solutions that have maximum advantages and lowest costs. Water conservation and recycling are two examples of environmental behaviours that the TPB has been successful in explaining. Furthermore, pro-environmental behaviour is motivated by moral obligations, understanding of the consequences of actions, and a sense of responsibility, according to the Norm-Activation Model (Schwartz, 1977) and the Value-Belief-Norm theory (Stern, 2000). The explanation of economical environmental intents and acts has been explained by these theories.

The more people who embrace sustainable behaviours, the more "normal" those behaviours become. People are more receptive to change in that direction when they observe an increasing number of persons around them adopting the lifestyles that society is trying to promote through policy changes, even if they do not personally live those lifestyles. (Minnesota Pollution Control Agency et al., 2009) thus, social acceptability significantly influences behavior, with societal trends often shaping individual actions.

Understanding the psychology of human behaviour is essential to promoting the adoption of sustainable technologies. Attitudes, values, and concerns significantly influence behavior, yet research reveals a weak correlation between what people believe and how they act. Because of this disconnect, taking sustainable action is not directly associated with merely believing in preserving the environment. Here's an in-depth look at ways to reduce this gap and encourage the use of sustainable technology.

1. The Role of Cognitive Dissonance

A key psychological theory in comprehending behaviour is "Cognitive Dissonance." This theory suggests that humans are driven to maintain consistency between their thoughts and actions. For instance, if someone acknowledges that saving energy is important for the environment, they may feel compelled to turn off lights or unplug devices when not in use, even if they previously didn't bother.

People are motivated to behave in ways that are consistent with what they believe by the discomfort that comes with cognitive dissonance.

2. Harnessing the Power of Social Norms

Human behaviour is frequently influenced by what other people do. In one study, signs in the rooms of the hotels encouraged customers to reuse their towels. While some banners only requested visitors to take into account the environment, others told them that most hotel guests had reused their towels at least once while they were there. The findings demonstrated that visitors were far more likely to reuse their towels if they're informed that others were doing so.

Participants in another study aimed at reducing household energy consumption received reports on their energy usage from researchers. Information regarding their energy consumption in comparison to their neighbours was included in certain reports. Individuals who obtained reports indicating that their energy usage exceeded that of their neighbours on average experienced a notable decrease in their energy usage. Thus, people often follow the actions of those they see in their peers, which is a reflection of the influence of social norms. People are more inclined to recycle if it becomes common in their society since they don't want to be the exception. Creating a societal norm around sustainable practices can be a potent force for change. This can begin small, in neighbourhoods, and then spread far beyond.

3. The Importance of Convenience

One of the main factors influencing people's adoption of sustainable behaviours is convenience. For example, most individuals are aware that they should ride their bike instead of driving to work or use reusable coffee cups, yet convenience frequently wins out. A new behaviour is less likely to be adopted the more work it needs.

Many people are aware of the advantages of using reusable shopping bags, but if they are hard to find or take extra work to carry, consumers may end up buying single-use plastic bags instead. Retailers can solve this by providing reusable bags at checkout stations or by clearly placing them close to the door. This facilitates consumers' decision to select a sustainable choice.

If there is no way to increase the advantages of a sustainable action, you can think about making the less sustainable choice more inconvenient. For instance, lowering the number of parking spaces available or increasing parking costs may deter people from driving to work and promote the use of public transportation or bicycles.

4. Education and Awareness

When knowledge is a barrier, educational strategies can be enhanced by using vivid imagery and breaking conventional thinking patterns. It's critical to customise your message to your target audience, making it actionable, relevant, and personal. Rather than making a broad appeal to "be more sustainable," for instance, it might be more successful to demonstrate how conserving water during droughts benefits the ecosystem and local community. As a result, the audience is more affected by the issue's urgency and relevance.

We can build an atmosphere where sustainable behaviours are not just possible but preferred by resolving cognitive dissonance, prioritising convenience, leveraging social norms, reducing barriers, fostering collaboration, and utilising targeted knowledge tactics. The potential lies in harnessing behavioral science to encourage the broad adoption of sustainable technologies, which will lead to significant and long-lasting change.

Conclusion

In conclusion, the future of sustainable technology adoption will be shaped by our ability to understand and influence human behavior. As new technologies emerge, they must be designed with consideration for psychological and social factors that drive adoption.

We must consider the intricate relationships that exist between humans and the environment, as these relationships are what give rise to environmental concerns. Finding leverage points to transition to more sustainable management can be facilitated by identifying important interactions. Recognising that humans are a component of ecosystems and that solving the problems would require teamwork is necessary. *"The greatest threat to our planet is the belief that someone else will save it,"* as Robert Swan once stated. That being said, every person has a responsibility to make informed decisions, and we all must contribute to preserving the planet's future by adopting sustainable technologies.

Also, it's important to understand that sustainable technology cannot be applied universally without adjustments. For ideas to be implemented successfully, they must be customised to the environmental and socioeconomic conditions of the local area. In order to effectively address environmental concerns in a holistic manner, collaboration across many stakeholders, including governments, corporations, academic institutions, and communities, is necessary for the adoption of sustainable technology.

The message is clear as we approach the nexus of environmental responsibility and technological advancement: supporting and embracing sustainable technology is not just an appealing option but a necessity. It's an act towards a sustainable and prosperous future for everybody, a promise to protect the planet's natural resources, and a commitment to future generations.

References

Ajzen, I. (1991). The theory of planned behavior. *Organizational Behavior and Human Decision Processes, 50*(2), 179–211. https://doi.org/10.1016/0749-5978(91)90020-t

Brandenburg, G., & Brandenburg, G. (2023, November 18). Sustainable Technology: Driving Environmental change in the Digital Age. https://ctrl-disrupt.nl/

en/-insights-news/sustainable-technology-driving-environmental-change-in-the-digital-age

Gardner, G. T., & Stern, P. C. (2002). Environmental problems and human behavior (2nd edition). Boston, MA: Pearson Custom Publishing.

Minnesota Pollution Control Agency, Manning, C., Elise Amel, & Su Beran. (2009). The Psychology of sustainable Behavior (T. Gaffey & S. Andre, Eds.) [Book].

Schwartz, S. H. (1977). Normative influences on altruism. In Advances in experimental social psychology (pp. 221–279). https://doi.org/10.1016/s0065-2601(08)60358-5

Stern, P. C. (2000). Toward a coherent theory of environmentally significant behavior. Journal of Social Issues, 56(3), pp. 407–424.

Webnx. (2023, July 10). Sustainable technology and its environmental impact: Paving the way for a greener future. Medium. https://medium.com/@webnxoffice/sustainable-technology-and-its-environmental-impact-paving-the-way-for-a-greener-future-260940f2bb8f

What is Sustainable Technology? | Rubicon. (n.d.). Rubicon. https://www.rubicon.com/sustainability-hub/articles/what-is-sustainable-technology/

Wilson, C. (2022, May 8). How can we change people's behaviour to act more sustainable? https://www.linkedin.com/pulse/how-can-we-change-peoples-behaviour-act-more-christina-wilson/

CHAPTER 17

PHYTOCHEMICAL ANALYSIS AND ANTIOXIDANT PROPERTIES OF PLEURITUS OSTREATUS

Dr. Bhawana Pandey[1] and Dr. Pratiksha Pandey[2]

[1]*Department of Biotechnology & Microbiology, Bhilai Mahila Mahavidyalaya, Bhilai*
[2]*Department of Botany, Bhilai Mahila Mahavidyalaya, Bhilai*
Email: bhawanapandey15@gmail.com

Abstract

The *Oyster mushroom* was cultivated in the Paddy straw at optimum conditions. The straw was kept in moisture and being kept in polythene bags and the temperature was maintained maximum at 25^{0}C. The oyster mushroom so cultivated was yellowish white in colour. The mushroom was dried under suitable conditions and was powdered, and the extraction was carried out as Soxhlet extraction using ethanol and water as solvents. Every extract was analysed for the qualitative phytochemical determination and antioxidant power determined using DPPH method. Different extracts show different phytochemical contents and antioxidant activity is taken into consideration the IC 50 Value for Ascorbic Acid is 28.90 µg/ml, Ethanol extract of Oyster mushroom is 43.40µg/ml and water extract Oyster mushroom 50.25 µg/ml. The water extract showed highest antioxidant power fallowed by ethanol extract.

Keywords: Soxhlet extraction, Phytochemical analysis, Ethanol extract

Introduction

Mushrooms are a group of fleshy macroscopic fungi, which recently, as other fungi were introduced into the plant kingdom of cell wall and spores. Mushroom has been valued and treated throughout the world as a special kind of food and medicine for thousands of years. There are many varieties of mushroom of which *pleurotus'* are characterized by a white spore print, attached to gills, often with an eccentric stipe or no stipe at all. They are commonly known as *"Oyster mushroom"*. Mushrooms are highly nutritious so they contain good quality proteins, vitamins and mineral. Mushrooms are low calorie food with little fat and are highly suitable for obese persons with no starch and very low sugars; they can serve as medicinal food for diabetic patient (Kirk *et al.*, 2008). Despite its nutritional value, mushroom cultivation is not widespread; many mushrooms are considered to be healthy food because they contain large enough protein needs of the rural poor especially during the rains (Cohen *et al.*,2002). It is also rich in some essential vitamins (B1, B2, C) and essential minerals than most plants. They also have a low fat content and hence high fiber content that enhances digestion. In this study, the mushroom, *Pleurotus ostreatus,* phytochemical analysis and antioxidant proprieties were done by the extraction using organic solvents such as methanol and water with the dried mushroom (Gregori *et al.*, 2007). Before extraction, the carbohydrate, protein and lipid content were estimated from the fresh mushroom.

This species is used as an edible mushroom and has various medicinal values. An antioxidant is a molecule that inhibits the oxidation of other molecules. Oxidation is a chemical reaction that can produce free radicals, leading to chain reactions that may damage cells. The free radicals can cause cancer, heart diseases, decline in brain function and immune system etc., Antioxidants terminates damage to the cells (Barrios *et al.*, 2009). The word pleurotus refers to 'sideways' which shows the sideways growth of the stem with respect to the cap. Oyster Mushroom (Sharma *et al.*, 2013). This is a common edible mushroom and it is also used industrially for mycoremediation purposes. The studies on phytochemical screening and antioxidant activity were defined by Patil *et al.* (2012). The *Pleurotus ostreatus* species were regarded as therapeutic food for their antioxidant, antimicrobial, antitumor and antiviral properties. These properties were obtained by the presence of various phytoconstituents, which includes terpenoids, tannins and carbohydrates in its various extracts (Hawksworth *et al.*, 2001). The antitumor activity was broadly explained by Hirano *et al.*(2001). Their study showed that the water-soluble polysaccharide was obtained from the fruiting bodies of *Pleurotus ostreatus* by hot water extraction and ethanol precipitation. In this study, the phytochemicals, antioxidant of the ethanol and aquous extracts of *Pleurotus ostreatus* (Bhatti *et al.*, 2007).

Oyster Mushroom

Phylum – Basidiomycotina

Class - Basidiomycetes

Sub class – Holobasidiomycetidae

Genus – *Pleurotus*

Species – *sajorcaju , sapidus , ostreatus , eous , membranaceous , florida , citrinopileatus, flabellatus .*

Origin

Cultivation of *Oyster mushroom* was started on test premise in Germany by Flack during the year 1917 on tree stumps and wood logs. Growing innovation was consummated in USA by Block, Tsao and Hau. The Latin pleurotus (sideways) alludes to the sideways development of the stem as for the cap, while the Latin ostreatus (and the English name, oyster) refers to the shape of the cap which resembles the bivalve of the same name. The name oyster mushroom is also applied to other *Pleurotus* species, so *P. ostreatus* is sometimes referred to as the tree oyster mushroom or the grey oyster mushroom to differentiate it from other species in the genus "*Pleurotus ostreatus*".

Morphology

Mushrooms have nutritive and utilized as human food from the time immortal. The mushroom has a broad, fan or oyster-shaped cap size 5–25 cm; colour ranges from white to grey or tan to dark-brown; the margin is smooth and often somewhat lobed or wavy. The flesh is white and varies in thickness due to stripe arrangement. The gills of the mushroom are white to cream. If so, the stipe is off-centre with a lateral attachment to wood. The spore of the mushroom is white to grey. The mushroom's stipe is often absent. When present, it is short and thick (Ali, 2007).

Different species of *Oyster mushroom*

- *Pleurotus ostreatus*
- *Pleurotus florida*
- *Pleurotus sajor-caju*
- *Pleurotus sapidus*
- *Pleurotus eous*
- *Pleurotus membranaceous*
- *Pleurotus flabellatu*

Importance of mushroom

Mushrooms are significant constituents of minor woods produce, that develop on the most plentiful biomolecule cellulose (Patel, 2012). Just fruiting body of the mushroom can be seen and other part is as mycelium (Wasser, 1999). Mushrooms have been discovered successful against cancer, cholesterol decrease, stress, insomnia, asthma, sensitivities and diabetes. (Medicinal value of edible fungi, 1983, pp. 203-209) Mushroom used as nutrient supplement for increasing immunity due to the presences of high protein. Even used for diabetic and heart patients due to the presences of low starch content. Organically dynamic mixes from the mushrooms have antifungal, antibacterial, cancer prevention agent and antiviral properties, and have been utilized as nematicides also.

Production Technology

The most suitable temperature for *Oyster mushroom* ranges from 20° to 30° C and humidity ranges from 55-70%. The best growing season for *Oyster mushroom* is the month of March/April to September/October. (Rajak, 2011).

Mushroom Cultivation

Oyster mushrooms are developed to its simple development innovation, accessibility of crude materials and number of species appropriate for climate (Khan, 2012). Falck, (1917) in Germany performed the first experimental cultivation of *P. ostreatus*. Kaufert (1935) reported medium, sexual spores of *Pleurotus carticatus* Fr. Furthermore; Block *et al.*, (1959) cultivated *P. ostreatus* first time under laboratory conditions using sawdust and oatmeal as substrate. In India (Bano, 1962) at CFTRI Mysore revealed development of *P. flabellatus* on paddy straw or corncobs as substrate were utilized under the sterile condition for developing *P. ostreatus* (Lindequist *et al.*, 2005) (Salmones *et al.*, 2005) fermentation of substrate used for the cultivation of *P.ostreatus* (Zadrazil, 1974) reported a method for continuous preparation of substrates used in *Pleurotus mushroom* (*P.ostreatus* and *P.florida*) cultivation. (Jandaik, 1974) grew, *P. sajor-caju* on various substrates including wheat and banana pseudostems. A method for cultivation of *P. sajor-caju* using cotton waste from textile industry (Singh, 1994).

Benefits of oyster mushrooms

Following benefits of oyster mushrooms include (Nanba,1993):

Inhibit free radicals:

- Pleuran, a polysaccharide found in oysters, is a potent antioxidant that helps protect the body's cells from free radical damage.
- Antioxidants found in oyster mushrooms aid in the fight against free radicals. Free radicals can cause heart disease and even cancer. As a result, it is extremely hazardous to the body. Furthermore, antioxidants have a significant function in immune system enhancement. They can aid to prevent the skin from premature aging.

Good for heart health:

- Oyster mushrooms include beta-glucan which helps reduce cholesterol levels in the blood while supporting heart health. Beta-glucan can alter cholesterol levels by absorbing cholesterol in the circulation.
- Oyster mushrooms include potassium, which helps keep the heart, neurons, and muscles working properly. Oyster mushrooms contain lovastatin, which helps manage cholesterol levels in the blood.

Good source of vitamin B3:

- Oyster mushrooms are high in vitamin B3, which is extremely good for the body. It is crucial in assisting the growth and function of the organs in the body. Vitamin B3 can aid in the breakdown of lipids, carbs, and proteins in your diet. As a result, they may be used to generate energy. When vitamin B3 enters the body, it is transformed into chemicals found in antioxidants. Hence, it can aid in the maintenance of a healthy physique.

May prevent diabetes:

- *Oyster mushrooms* help manage blood sugar levels. It, therefore, lowers the likelihood of developing type II diabetes.
- Oyster mushrooms include vitamin B, which helps the digestive system and red blood cells function better.

Help prevent cancer:

- Oyster mushrooms inhibit the development of colon cancer and breast cancer cells.
- They act to keep cancer cells from spreading to other places of the body. The glycoprotein present in oyster mushrooms boosts the body's anti-cancer response naturally. The specific pleuran polysaccharide combats colon cancer. Oyster mushrooms help prevent cancer by reducing DNA damage.

Good source of vitamin D:

- Vitamin D or cholecalciferol content is very high in oyster mushrooms because it contains about 25 µg in one serving of 86 grams, which fulfills about 166.67 percent of our daily requirement of Vitamin D.
- Vitamin D is required by the human body to maintain the balance of insulin in our bodies. Calcium is required for the efficient release of the insulin hormone.

Aids in the development of a strong and healthy immune system:

- Oyster mushrooms are high in minerals and vitamins B3, B5, B6, D, and A, which are known to be essential for leucocyte functioning.
- They are high in phosphorus, magnesium, calcium, salt, iron, and zinc, all of which are necessary for the human body to develop robust immunity.

Good for neurological health:

- Dopamine and serotonin are important for maintaining mental wellness, and the copper content of oyster mushrooms allows us to do so.
- Oyster mushrooms contain folic acid, a B group vitamin that promotes good sleep and memory retention and may even aid with mood regulation.

Aid in the treatment of anemia:

- Oyster mushrooms include a variety of vitamins and minerals that help prevent anemia by stimulating the body's blood production mechanism. Vitamin B2 is one such vitamin that is required for the normal production of hormones. Vitamin B2 keeps oxygen moving throughout our bodies and allows iron to be mobilized.

Good for teeth:

- Calcium is essential for oral health and strength. Calcium, phosphorus, and vitamin D are beneficial for maintaining bone density in general, particularly jawbone density, which is responsible for holding our dental structure together.
- Oyster mushrooms have high levels of these three nutrients, which improve dental health by strengthening tooth enamel.

Good for dieters:

- Oyster mushrooms are the ideal diet food because they are low in calories, fat-free, and high in fiber, which helps keep you satisfied for a long period.

Anti-inflammatory properties:

- Oyster mushrooms have potent anti-inflammatory effects. According to research, they may diminish the release of a variety of inflammatory indicators in the body.
- This might have far-reaching implications because reducing inflammation may help bring treatment for a variety of inflammatory illnesses ranging from rheumatoid arthritis to inflammatory bowel disease.

Holistic nutrition during pregnancy:

- When you are pregnant, your nutritional demands increase. In case, oyster mushrooms may be a preferable choice.
- Riboflavin, potassium, vitamin B6, folic acid (folate), magnesium, vitamin C, pantothenic acid, and amino acids are present in oyster mushrooms.

Scope of the Study

This study is based on the edible mushrooms which are known to be rich in high quality protein, high proportion of unsaturated fatty acids and have vitamins and minerals content. Various compounds of medicinal value present in mushroom are used as neutraceutical and dietary supplements. In this study the edible mushroom *Agaricus bisporus* and *Pleurotus ostreatus* was cultivated in agricultural waste using certain chemicals and optimum conditions for mass production. Mushroom cultivation gives economic help to the cultivators and its consumption is good for health too. Mushroom cultivation is a very easy process and they are eco-friendly and healthy in nature as well. They contribute to nature as a natural decomposer which reduces the amount of agricultural waste by decomposing it and releasing the nutrients that are entrapped inside them.

Objectives of the study are mentioned below

1. Collection and identification of *Oyester mushrooms.*
2. Analysis and screening of Phytonutrients and Phyochemicals.
3. Analysis of antioxidant properties of *Oyester mushrooms.*

Material and Methods

A. Preliminary Phytonutrients Analysis:

- **Estimation of Carbohydrates by Anthrone Method**: The carbohydrate content was estimated with anthrone reagent as suggested by Yemm & Willis (1954).
- **Estimation of Protein by Folin-Ciocalteau's:** Protein content in the sample was estimated by the Lowry method (1951) using Folin-Ciocalteau's Reagent.
- **Extraction of Lipid:** The total lipid content in the sample was extracted by the method followed from Ryckebosch *et al.* (2012).

B. Preliminary Phytochemical Analysis (Rajewska *et al.*, 2005):-

- **Test for steroids**: Two ml of acetic anhydride was added to 0.5 g ethanolic extract of each sample with 2 ml H_2S0_4. The colour changed from violet to blue or green in some samples indicating the presence of steroids.
- **Test for terpenoids (Salkowski test)**: Five ml of each extract was mixed in 2 ml of chloroform, and concentrated H_2S0_4 (3 ml) was carefully added to form a layer. A reddish brown colouration of the inter face was formed to show positive results for the presence of terpenoids.
- **Test for cardiac glycosides (Keller-Killani test)**: Five ml of each extracts was treated with 2 ml of glacial acetic acid containing one drop of ferric chloride solution. This was under layered with 1 ml of concentrated sulphuric acid. A brown ring of the interface indicates a deoxy sugar characteristic of cardenolides. A violet ring may appear below the brownring, while in the acetic acid layer, a greenish ring may form just gradually throughout thin layer.
- **Test for alkaloids**

  Mayer's test : To a few ml of the filtrates, a drop of Mayer's reagent was added by the side of the test tube. A creamy or white precipitate indicates the test is positive.

  Hager's test: To a few ml of filtrate, 1 or 2ml of Hager's reagent were added. A prominent yellow precipitate indicated the test as positive.

- **Test for phenolic compounds**

  Ferric chloride test: The extract was diluted to 5 ml with distilled water. To this a few drops of neutral 5% ferric chloride solution was added. A dark green color indicates the presence of phenolic compounds.

  Test for tannins: About 0.5 mg of dried powdered samples was boiled in 20 ml of water in test tubes then filtered. A few drops of 0.1 % ferric chloride was added and observed for brownish green or blue black coloration.

  Test for flavonoids: To 5 ml of the dilute ammonia solution a portion of the aqueous extract was added, followed by addition of concentrated sulphuric acid. Appearance of yellow coloration indicates the presence of flavonoids.

  Test for saponins: The extract was diluted with distilled water and made up to 20 ml. The suspension was shaken in a graduated cylinder for 15 min. 2 cm layer of foam indicates the presence of saponins.

C. Antioxidant Properties of *Oyester mushrooms*

- **DPPH Scavenging Activity**: The ability of the plant extract to scavenge 1,1- diphenyl-2-picryhydrazyl (DPPH) free radicals was assessed by the standard method . The stock solution of extract was prepared in ethanol to achieve the concentration of 1 mg/ml. Dilutions were made to obtain concentrations of , 10, 20, 30, 40, 50, 60, 70, 80, 90 and 100 µg/ml. Diluted solutions (1 ml each) were mixed with 3 ml of ethanolic solution of DPPH (DPPH, 0.004%). After 30 min of incubation at room temperature the reduction of the DPPH free radical was measured by reading the absorbance at 517nm using UV-Visible Spectrophotometer. Initially, absorption of blank sample containing the same amount of ethanol and DPPH solution was prepared and measured as control. Ascorbic acid was used as standard. The experiment was carried out in triplicate. Percentage inhibition was calculated using equation, while IC50 values were estimated from the % inhibition versus concentration plot, using a non-linear regression algorithm.

  (Absorbance of control- absorbance of sample)

  % inhibition= ---------------------X 100 equation-- (1) Absorbance of control

Result and Discussion

Preliminray phytochemical Analysis: The phytochemical analysis of for Oyster mushroom was conducted for water and Ethanol extract. The results showed that the mushroom is rich in almost all types of secondary metabolites which are essential for life.

In-Vitro Antioxidant Activity by DPPH Scavenging Method: The mushroom was dried under suitable conditions and was powdered, and the extraction was carried out as Soxhlet extraction using ethanol and water as solvents. Every extract was analysed for the qualitative phytochemical determination and antioxidant power determined using DPPH method. Different extracts show different phytochemical contents and as per the antioxidant power is taken into consideration the IC 50 Value for Ascorbic Acid is 28.90 µg/ml, Ethanol extract of *Oyester mushrooms* is 43.40 µg/ml and water extract *Oyester mushrooms* 50.25 µg/ml. The water extract showed highest antioxidant power fallowed by ethanol extract.

Plant sample: – The *Oyster* mushroom was collected from Durg district of C.G It was then identified.

Fig. 1-4: Cultivation of *Oyester Mushroom*

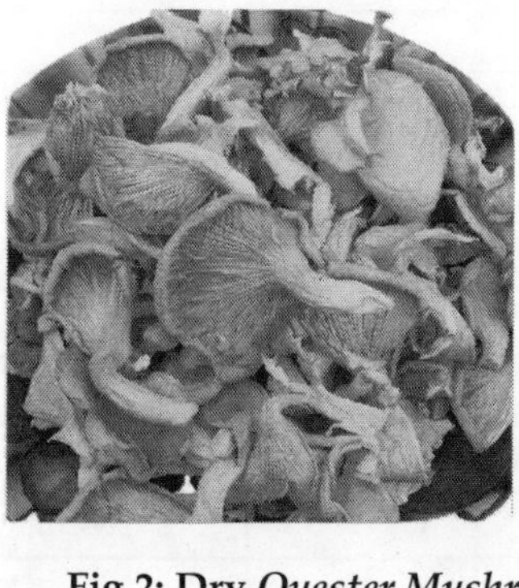

Fig 2: Dry *Oyester Mushroom*

Fig 3: Powder of *Oyester Mushroom*

Sample preparation

The *P. ostreatus* sample was thoroughly washed to remove impurities and water residues were removed from its surface with a sterile towel. The sample was then dried in an oven with hot air at 55–65 C and pulverised with a blender. The resulting powder was weighed at room temperature and stored in airtight containers for subsequent use.

Processing

The mushroom was dried in shade and pulverized into fine powder by using mortal pistol.

Preparation of the extract :- 10 gram powder of oyster mushroom weighed and transferred into 10 ml conical flask 100 ml of ethanol was added in flask. The conical flask was closed by foil paper and placed at room temperature for 72 hours and crude extracts was then filtered through whatmann no. 1 filter paper.

Table 1: Phytochemical analysis of *Pleurotus ostreatus*

SN	Phyto constituents	Ethanol Extract	Aqueous Extract
1	Carbohydrates	-	+
2	Tannins	+	+
3	Saponins	+	+
4	Flavonoids	-	-
5	Alkaloids	-	-
6	Quinones	+	+
7	Glycosides	-	-
8	Cardiac glycosides	+	-
9	Terpenoids	-	+
10	Phenols	+	+
11	Coumarins	-	-
12	Steroids	-	+

Table 2: Phytonutrients and Minerals of *Oyster Mushroom*

SN	Nutrient	Quantity	S.N.	Minerals	Quantity
1.	Water	76.69gm	1.	Ca	8mg
2.	Protein	2.85g	2.	Fe	1.14mg
3.	Lipid (fat)	0.35g	3.	Mg	15mg
4.	Ash	0.85g	4.	P	103mg
5.	Carbohydrate	5.24g	5.	K	361mg
6.	Energy	2.0g	6.	Na	15mg
7.	Sugar	095g	7.	Zn	0.66mg

SN	Nutrient	Quantity	S.N.	Minerals	Quantity
			8.	Cu	0.210mg
			9	Mn	0.097mg
			10.	Se	2.2mg

Table 3: Phytochemical Test results for *Oyster Mushroom*

Phytochemical	Observation	Result
Alkaloid test	Formation of reddish brown precipitate	+ve
Steroid	Formation of persistent foam	-ve
Flavonoid	Formation of yellow colour and colourless addition of HCL	-ve
Tannins	Formation of blue greenish	-ve
Saponin	Formation of persistent foam	+ve
Terpenoid	Formation of reddish brown colour	+ve
Phenol	Deep blue or black colour	-ve
Glycoside	Brown ring at the interface below greenish ring	+ve
Carbohydrate	Red or dull violet colour	+ve
Quinine	Yellow precipitate	+ve
Resin	Turbidity occur	+ve
Coumarine	Formation of yellow colour	+ve

In-Vitro Antioxidant Activity by DPPH Scavenging Method

The invitro antioxidant properties of *Oyster mushroom* for two extracts namely Ethanol and water were done taking with Ascorbic Acid as standard and results are shown in Table-4 -6 to and plotted in Graph-1-3. The IC50 Value for Ascorbic Acid is 28.90 µg/ml, Ethanol extract of *Oyster mushroom* is 43.40 µg/ml and water extract *Oyster mushroom* 50.25 µg/ml.

Table 4: DPPH Scavenging activity of standard Ascorbic acid

SN	Conc.	Abs. of Ascorbic Acid	% reduction	IC_{50} Value
1	10	0.490	39.60	28.90
2	20	0.290	43.80	
3	30	0.263	50.20	
4	40	0.220	53.35	
5	50	0.199	59.12	
6	60	0.180	61.20	
7	70	0.159	65.20	
8	80	0.138	70.40	
9	90	0.122	74.20	
10	100	0.085	80.11	

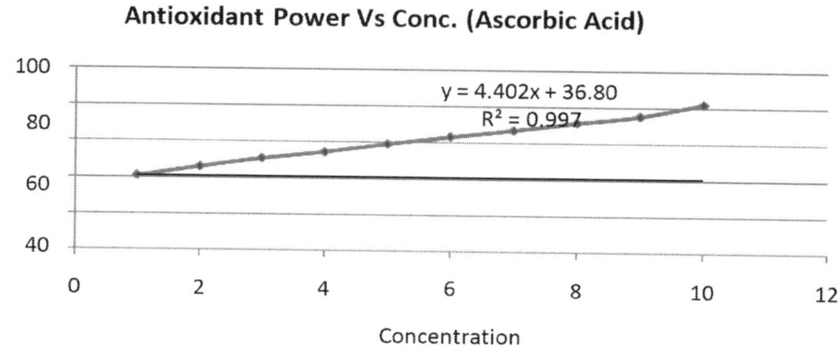

Graph-1: DPPH Scavenging activity of standard Ascorbic acid

Table-5: DPPH Scavenging activity Ethanol Extract of *Oyster mushroom*

SN	Concentration In (µg/ml)	Abs. of Ascorbic Acid	% reduction	IC_{50} Value
	Control	0.490		
1	10	0.291	40.60	
2	20	0.282	42.80	
3	30	0.270	44.40	
4	40	0.262	46.35	
5	50	0.236	53.16	43.40
6	60	0.220	56.20	
7	70	0.198	58.70	
8	80	0.179	65.40	
9	90	0.170	64.50	
10	100	0.158	68.31	

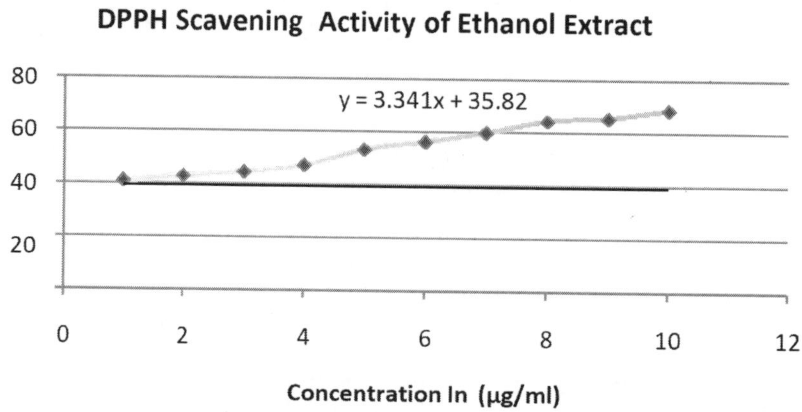

Graph 2: DPPH Scavenging activity of Ethanol Extract of *Oyster mushroom*

Table-6: DPPH Scavenging activity of water Extract of *Oyster mushroom*

S.No.	Conc (µg/ml)	Absorbance	% Reduction	IC_{50} Value (µg/ml)
	Control	0.490	-	
1.	10	0.360	21.48	
2.	20	0.338	26.36	
3.	30	0.312	33.62	
4.	40	0.278	43.52	50.25
5.	50	0.238	51.78	
6.	60	0.213	54.68	
7.	70	0.186	62.32	
8.	80	0.158	71.58	
9.	90	0.114	76.45	
10.	100	0.092	82.65	

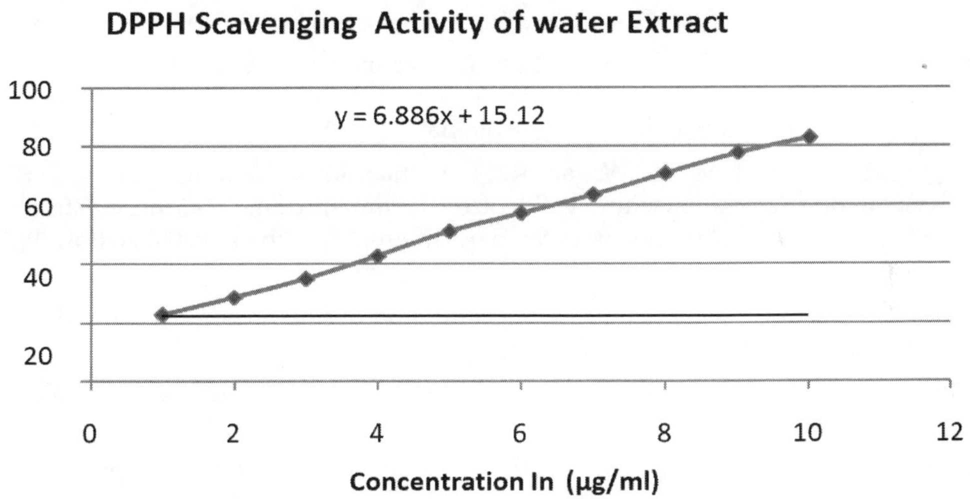

Graph 3: DPPH Scavenging activity of water Extract of *Oyster mushroom*

Discussion and Conclusion

The mushroom was cultivated in paddy straw. As per the phytochemical analysis showed that the mushroom is rich in almost all types of secondary metabolites which are essential for life, these phytochemicals boost up the all biological systems. So as per the recommendations the mushrooms should be in taken in daily life. The other property was the antioxidant property of this mushroom. The mushroom showed a comprehensive antioxidant property. All types of diseases are due of the accumulation of free radicals in the body of humans. Almost all types of diseases do get originated due to the excessive deposition of free radicals. In order to get healthier these free radicals showed get washed out from the body. There is only way to stop this process is the interaction between two free radicals because a radical is a species having only an unpaired electron. So the neutralization of free radicals is compulsory which is possible either due to the donation or acceptance of that electron. To overcome this problem there are some aromatic phytochemicals in the form of phenols mostly present in the plants. As per these properties of mushroom is taken into consideration it showed a healthy response as an anti oxidant, so consumption of mushroom is useful for a healthy life.

References

Ali, M. A., Mehmood, M. I., Nawaz, R., Hanif, M. A., & Wasim, R. (2007). Influence of substrate pasteurization methods on the yield of *oyster mushroom* (*Pleurotus* species). Pak. J. Agric. Sci, 44, 300-303.

Bano Zakia, and Rajarathnam, S. 1982. Pleurotus mushroom as a nutritious food. In: Tropical Mushroom - Biological Nature and Cultivation Methods. Chang, S.T. and Quimio, T.H. (Eds.). The Chinese University Press, Hong Kong, 363 pp.

Barrios Espinoza, B. M., Moreno Ruiz, L., & Sánchez, J. E. (2009). Composteo encajones de Maderacomopretratamiento del sustrato para cultivar Pleurotus ostreatus. Revista mexicana de micología, 29, 51-59.

Bhatti, M. I., Jiskani, M. M., Wagan, K. H., Pathan, M. A., & Magsi, M. R. (2007). Growth, development and yield of oyster mushroom, Pleurotus ostreatus (JACQ. EX. FR.) Kummer as affected by different spawn rates. Pak. J. Bot, 39(7), 2685-2692.

Block, S.S, Tsao, G. and Han, L. 1958. Production of mushroom from sawdust. J. Agric. Food Chem., 6: 923-927.

Chang, S.T., 2006. The world mushroom industry: World J. Fungal & Plant Biol., 3 (1): 01-12, 2012 7

Cohen, R., L. Persky and Y. Hadar, 2002. Biotechnical applications and potential of wood-grading mushrooms of the genus Pleurotus. Appl. Microbiol. Biotechnol., 58: 582-594.

Gregori, A., M. Svagelj and J. Pohleven, 2007. Cultivation techniques and medicinal properties of Pleurotus spp. Food Technol. Biotech., 45: 238-249.

Hawksworth, D.L., 2001. Mushrooms: the extent of the unexplored potential. Int. J. Med. Mush., 3: 333-340.

Hirano, R., Sasamoto, W., Matsumoto, A., Itakura, H., Igarashi, O., & Kondo, K. (2001). Antioxidant ability of various flavonoids against DPPH radicals and LDL oxidation. Journal of nutritional science and vitaminology, 47(5), 357-362.

Jandaik, C.L. and Kapoor, J.N. 1974. Studies on the cultivation of Pleurotus sajorcaju (Fr.) Singer. Mushroom Sci., 9: 667-672.

Kaufert, F. 1935. The production of asexual spores by Pleurotus corticathus. Mycologia, 27: 333-340. K

Khan, M.A. and M. Tania, 2012. Nutritional and medicinal importance of Pleurotus mushroom: An overview. Food Review Int., 28(3): 313-329.

Kirk, P.M., P.F. Cannon, D.W. Minter and J.A. Stalpers, 2008. Dictionary of the Fungi (10 ed.), th CABI, Wallingford, UK.

Lindequist, U., T.H.J. Niedermeyer and W.D. Julich, 2005. The pharmacological potential of mushrooms: Evidence based complement. Altern. Med., 2(3): 285-299.

Lowry, O. H., Rosebrough, N. J., Farr, A. L., and Randall, R. J. (1951) Protein measurement with the Folin phenol reagent. J. Biol. Chem.

Nanba, H., 1993. Maitake mushroom-the king mushroom. Mush. News, 41(2): 22-25.

Patel, Y., Naraian, R., & Singh, V. K. (2012). Medicinal properties of Pleurotus species (Oyster mushroom): A review. World Journal of Fungal and Plant Biology, 3(1), 1-12.

Patil, S. S. (2012). Cultivation of Pleurotus sajorcaju on different agro wastes. Science Research Reporter, 2(3), 225-228.

Rajak, S., Mahapatra, S. C., & Basu, M. (2011). Yield, Fruit Body Diameter and Cropping Duration of Oyster Mushroom (Pleurotus sajorcaju) Grown on Different Grasses and Paddy Straw as Substrates. European Jr. of Medicinal Plants, 1(1), 10-17.

Rajewska, J. and B. Balasinska, 2004. Biologically active compounds of edible mushrooms and their beneficial impact on health. Postepy. Med. Dosw., 58: 352-357.

Ryckebosch, E., Muylaert, K. and Foubert, I. (2012) Optimization of an Analytical Procedure for Extraction of Lipids from Microalgae. Journal of the American Oil Chemists' Society, 89, 189-198. https://doi.org/10.1007/s11746-011-1903-z

Salmones, D., G. Mata, K.N. Waliszewski, 2005. Comparative culturing of Pleurotus spp. biomass production and substrate biodegradation. Bio. Technol., 96(5): 537-544.

Sharma, S., Kailash, R., Yadav, P., &Pokhre, C. P. (2013). Growth and Yield of Oyster mushroom (Pleurotus ostreatus) on different substrates. J. Vol. l Rep, 2(1), 03-08.

Singh, C.S., R. Sharma, Y. Patel and V.K. Singh, 2010. Management of Agricultural Solid Wastes by Locally Isolated Pleurotus Species, a White Rot Fungi. Asian J. Microbiol. Biotech. Env. Sci., 13: 561-564

Wasser, S.P. and A.L. Weis, 1999. Medicinal properties of substances occurring in higher Basidiomycetes mushrooms: Current perspectives (Review). Int. J. Med. Mush., 1: 31-62.

Wasser, S.P., 2002. Medicinal mushrooms as a source of antitumor and immune modulating polysaccharides (minireview). Appl. Microbiol. Biotechnol., 60: 258-274.

Yemm, E. and Willis, A.J. (1954) The estimation of carbohydrate in plant extracts by Anthrone. Biochemical Journal, 57, 508-514.

Zadrazil F., 1978. Cultivation of Pleurotus. In: The Biology and Cultivation of edible mushrooms, Eds., S.T. Chang and W.A. Hayes, Academic Press, USA, pp: 521-557.

CHAPTER 18

MICROALGAE ARE FUTURE RESOURCE OF BIOFUEL

Dr. Indu Soni[1] Madhu Yamini[2] and Dr. Suneeta Patra[3]

[1]*Department of Botany, B. P. L. P. G. Govt. College Arang Raipur (C.G.)*
[2]*Department of Botany Atal University Bilaspur (C.G.)*
[3]*Department of Botany Govt. N.P.G. Science College Raipur (C.G.)*

Abstract

The majority of algae that are intentionally cultivated fall into the category of microalgae . A micro alga is an ideal organism for biofuels. It can be cultivated on non-arable land and it grows in ponds or photo-bioreactors. Microalgae are capable of producing large amounts of biomass and usable oil in either high rate algal ponds or photobioreactor. This oil can then be turned into biodiesel which could be sold for use in automobiles. The species of chlorella and spirulena is more important for biofuel. The present study can give information's on bio-fuel which produce from microalgae. Advantages of microalgae from biodiesel is –high productivity and contain high oil content compared with terrestrial crops, up to 77 wt% lipid content, negligible lignocellulosic biomass component, diesel more versatile and energy dense than bio-ethanol. Terrestrial oil crops are unable to meet in future bio-oil demand but by using of micro-algae it is possible.

Key words: Micro-algae, biofuel, photobioreactor, spirulina, biomass.

Introduction

Algal Culture

Algal culture is a form of aquaculture involving the farming of species of algae. The majority of algae that are intentionally cultivated fall into the category of microalgae (also referred to as phytoplankton microphytes or planktonic algae).

Macroalgae, commonly known as seaweed, also have many commercial and industrial uses, but due to their size and the specific requirements of the environment in which they need to grow, they do not lend themselves as readily to cultivation (this may change, however, with the advent of UAS seaweed cultivators using up flowing air bubble in small containers.

Material and Method

Growing &, Harvesting Of Algae

Growing algae-Water, carbon dioxide, minerals and light are all important factors in cultivation, and different algae have different requirements. The basic reaction for algae growth in water is carbon dioxide + light energy + water = glucose + oxygen + water. This is called *autotrophic* growth.

It is also possible to grow algae in darkness, using sugars (such as glucose) to feed the algae directly, in which case they do not use carbon dioxide. This is called *heterotrophic* growth.

The following factors are responsible for growing algae-

1. Temperature

The water must be in a temperature range that will support the specific algal species being grown mostly between 25 to 35 degrees.

2. Light and mixing

In a typical algal-cultivation system, such as an open pond, light only penetrates the top 3 to 4 inches (76–100 mm) of the water, though this depends on the algae density. As the algae grow and multiply, the culture becomes so dense that it blocks light from reaching deeper into the water. Direct sunlight is too strong for most algae, which can use only about $^{1}{}_{10}$ the amount of light they receive from direct sunlight; however, exposing an algae culture to direct sunlight (rather than shading it) is often the best course for strong growth, as the algae underneath the surface get more light.

3. Odor and oxygen

The odor associated with bogs, swamps, indeed any stagnant waters, can be due to oxygen depletion caused by the decay of deceased algal blooms. Under anoxic conditions, the bacteria inhabiting algae cultures break down the organic material and produce hydrogen HYPERLINK "http://en.wikipedia.org/wiki/Hydrogen_sulfide"sulfide and ammonia which causes the odor. This hypoxia often results in the death of aquatic animals. In a system where algae are intentionally cultivated, maintained, and harvested, neither eutrophication nor hypoxia are likely to occur.

Some living algae also produce odorous chemicals, particularly certain blue-green algae (cynobacteria) such as *Anabaena*. The most well-known of these odor-causing chemicals are MIB (2-methylisoborneol) and geosmin. They give a musty or earthy odor that can be quite strong. Eventual death of the cynobacteria releases additional gas that is trapped in the cells. These chemicals are detectable at very low levels, in the parts per billion range, and are responsible for many "taste and odor" issues in drinking water treatment and distribution. Cynobacteria can also produce chemical toxins that have been a problem in drinking water.

4. Nutrients

Nutrients such as nitrogen (N), phosphorus (P), and potassium (K) serve as fertilizer for algae, and are generally necessary for growth. Silica and iron, as well as several trace elements, may also be considered important marine nutrients as the lack of one can limit the growth of, or productivity in, a given area. These elements must be dissolved into the water, in bio-available forms, for algae to grow.

Fig 1: Algal cell culture growing

Pond and Bioreactor Cultivation Methods

Algae can be cultured in open-ponds and photo bioreactors

1. Open Pond

Ponds and lakes are open to the elements. Open ponds are highly vulnerable to contamination by other microorganisms, such as other algal species or bacteria. Thus cultivators usually choose closed systems for monocultures. Open systems also do not offer control over temperature and lighting. The growing season is largely dependent on location and, aside from tropical areas, is limited to the warmer months.

2. Photobioreactors

Algae can also be grown in a photo bioreactor (PBR). A PBR is a bioreactor which incorporates a light source. Virtually any translucent container could be called a PBR; however the term is more commonly used to define a closed system, as opposed to an open tank or pond. Because PBR systems are closed, the cultivator must provide all nutrients, including CO2.

A PBR can operate in "batch mode», which involves restocking the reactor after each harvest, but it is also possible to grow and harvest continuously. Continuous operation requires precise control of all elements to prevent immediate collapse. The grower provides sterilized water, nutrients, air, and carbon dioxide at the correct rates. This allows the reactor to operate for long periods. An advantage is that algae that grows in the «log phase" is generally of higher nutrient content than old "senescent" algae. Algal culture is the culturing of algae in ponds or other resources.

Fig 2: Turbulent thin-film photobioreactor

Different types of PBRs include:

- Polyethylene sleeves or bags.
- Tank.
- Glass or plastic tubes.

Harvesting

Algae can be harvested using micro-screens, by centrifugation, by flocculation and by froth flotation.

Interrupting the *carbon dioxide* supply can cause algae to flocculate on its own, which is Called "auto flocculation"

Harvesting Biodiesel

- **Micro algae have much faster growth-rates than terrestrial crops.**
- **Algal-oil processes into biodiesel as easily as oil derived from land-based crops.**
- **Use micro algae**
  - Less complex structure
  - Faster growing rate
  - High oil content
- **How to harvest**
  - Open-pond systems
  - Type of algae has to be hardy
  - Can be less hardy and grow slower

- **Use Bioreactor Tubes**
- **Use existing infrastructures**

  - Provides the raw materials for the system, such as CO2 and nutrients

  - Changes those wastes into resources.

Types of Cultures

1. **Clonal culture** - When a part of algae or unicellular algae in culture position. This called clonal culture.
2. **Axenic culture** -A specific type of algae strength is called pure or axenic culture.
3. **Unialgal culture** - Separation of specific algae and than growing up by perfect media are called unialgal culture.
4. **Synchronous culture** - It is controlling the system of neutration, temperature and light by which the measurement growing rate and population development.
5. **Enrichment culture** - Obtain from some nutrition elements which are created to development by single or uni-algal that called enrichment culture.

Fig 3: Algal culture collection

Specific algal strains can be acquired from algal culture collections, with over 500 culture collections registered with the World Federation for Culture Collections.

Algal Culture Collection Application

Algal culture is importance from following reason:-

1. To study the life cycle and morphology of specific algae.
2. To study the physiology, biochemical, genetic, and ultra structure of algae.
3. To study the various stage of reproduction like gamete formation, gamete fusion and related factor.
4. To study the productivity rate of some specific algae and their contact from immunology.
5. To determine the chemical composition of algae.

Importance of Algae

- Algae is easy to grow.
- Can produce a high yield of oil.
- Oil can be processed into biodiesel.
  - Help to solve dependence on fossil fuels.
  - Can be better for the Earth.

Use of algae

1. **Food-** Several species of algae are raised for food.
2. **Fertilizer and agar** - For century's seaweed has been used as fertilizer it is also an excellent source of potassium for manufacture of potassium nitrate.
3. **Pollution control** - It is help to control pollution.
4. **Energy production -** Business, academia and governments are exploring the possibility of using algae to make gasoline, diesel and other fuels.
5. **As fodder –** Algae is use as fodder of animal.

Other Use

- **Hydrogen - Algae can be grown to produce hydrogen**
- Discovered first in 1939 by Hans Gaffrom
- Late 1990's it was found that if sulfur deprived, algae will produce hydrogen
- **Biomass - Algae can be grown to produce biomass**
- Burned to produce heat and electricity
- Can still produce greenhouse gases

Conclusion

Alage, including seaweed, are potential sources of renewable fuel, food and chemicals. This post note examines the technical, economic and environment issue around algal bio-fuels and their relevance to UK and EU policy target. Effluent from algal bio-fuel production must meet water quality standards, and producers require permits under conditions set out in the Environmental Permitting Regulations 2010 (EPR). Possible adverse environmental impacts include algal blooms or contamination by heavy metals. These may be accumulated by algae if they are present in seawater, waste water or flue gas.

The process of generating biofuel from algae involves the growth, concentration, separation and conversion of microalgae biomass, some of which can be genetically altered. After separating biomass, a significant portion of by product remains. It is important that the remaining by product have a useful and safe purpose for the economic feasibility and environmental sustainability of the process

References

Cultivating Algae for Liquid Fuel Production (*http://oakhavenpc.org/cultivating_algae.htm*); NREL, 2005

Department of Energy, Office of Fuel Development. "Aquatic Species Program". 1996.

Enhanced Biofuels & Technologies Ltd. 2007. Accessed: http://www.ebtplc.com/c4c.htm

Guiry, M.D. and Blunden, G. (Eds) 1991. Seaweed Resources in Europe: Uses and Potential. John Wiley & Sons. ISBN 0-471-92947-6

Murnford, T.F. and Miura, A. 1988. 4. Porphyra as food: cultivation and economics. p.87 — 117. In Lembi, C.A. and Waaland, J.R. (Ed.) Algae and Human Affairs. 1988. Cambridge University Press. ISBN 0 521 32115 8

John Sheehan, Terri Dunahay, John Benemann and Paul Roessler, "A Look Back at the U.S. Department of Energy's Aquatic Species Program-Bio-diesel from Algae, Closeout Report", July 1998, NREL/TP-580-24 190 http://www.nrel.gov/docs/legosti/fy98/24190.pdf

Michael Briggs, Widescale Biodiesel Production from Algae, University of New Hampshire, Physics Department, revised August 2004. http://www.unh.edu/p2/biodiesel/article_alge.html

Sheehan, J., T. Dunahay, J. Benemann, and P. Roessler. 1998. A look back at the U.S. Department of Energy's aquatic species program - Biodiesel from algae. US Dept. Energy, Office of Fuels Development, Nat. Renewable Energy Lab., Golden, CO.

Solix BioFuels, 2006. accessed: http://www.solixbiofuels.com/index.php?option=com_content&task=view&id=12&Itemid=26

http://www.ecology.com/dr-jacks-natural-world/most-important-organism/index.html

http://journeytoforever.org/biodiesel_yield.html

CHAPTER 19

REMOTE SENSING: APPLICATION IN AGRICULTURE

Dr. Veena Thakur[1] & Dr. Rameshwar Prasad Thakur[2]

[1]Assistant Professor Botany, Govt Pt. Shyam Shankar Mishra College Gariyaband, C.G.

[2]Assistant Professor Geography, Madan Lal Sahu Govt. College Armarikala, C.G.

vinathakur@gmail.com

Abstract

Humanity depends on agriculture for food, fiber, fuel, and raw materials, all of which are essential to survival. In today's world, this responsibility needs to be fulfilled in the context of environmental sustainability, climate change, and an enormous and continuously growing human population, all the while preserving the viability of agricultural activities to guarantee livelihoods and subsistence. Remote sensing can help agricultural methods adapt and evolve to meet this significant issue by giving consistent information on crop status at various scales and for various players throughout the growing season. Through the provision of creative solutions to enhance crop management, track environmental conditions, and maximize resource use, remote sensing technology has shown to be a transformative force. A new era of precision agriculture and sustainable resource management could be ushered in by integrating remote sensing technology into agricultural methods. Present chapter deals with various applications of remote sensing in the field of agriculture to improve the living standards.

Key words: Agriculture, Crop, Environment sustainability, Remote sensing

Introduction

An important factor in India's economic expansion is agriculture. Agriculture and related sectors employ approximately 54.6% of the labour force and generate 17.8% of the nation's Gross Value Added (GVA). Global food security is largely dependent on rice, the world's most significant grain. Rice production in Asia leads the world in this regard. Food security worldwide is now much more threatened as a result of Asia's limited arable land area and the region's mounting food demands brought on by the region's extraordinary population growth.

Food production must rise in response to the growing global population, which presents a challenge to civilization. Sustainable management of the whole agricultural system is required to increase food output. Limited production regions necessitate agricultural intensification. In order to do this, new agricultural technology and management techniques that lessen the negative impacts of greater input use—such as pesticides and fertilizers—must be implemented. India has historically had low agricultural production due to the adoption of conventional farming methods, which have historically resulted in considerably less improvement in efficiency and agricultural output. In response to this, the government launched the fourth wave of the agricultural revolution, bringing new technologies to the field and raising yields. It is believed that India's agriculture industry is currently poised for a revolutionary technological shift. Precision agriculture is considered to have arrived as a result of disruptive technologies, following decades of progress that began with mechanization and the Green Revolution. The new farm management strategy ensures that individual fields or crops receive precisely the inputs they need for maximum yield by precisely mapping farmlands using Geopolitical Systems (GPS) and Artificial Intelligence-enabled software.

In today's agriculture, remote sensing technologies are crucial, greatly increasing productivity and sustainability. The field of agriculture can make use of a variety of remote sensing technology. This covers a variety of imaging techniques such as radar, thermal imaging, multispectral imaging, light detection and ranging (LiDAR), aerial and satellite imagery, unmanned aerial vehicles (UAVs), etc. Sensor-equipped drones and satellites greatly improve agricultural productivity and decision-making processes, enabling data-driven insights to solve a range of issues. In the field of agriculture, various kinds of remote sensing sensors can be used for particular purposes. Many benefits can result from the use of remote sensing in agriculture, including accurate management of nutrients, effective models for managing drought stress, efficient models for modeling water consumption, mapping yield, estimating of yield, and crop phonological phases. The soil and crop mapping processes can be carried out through image processing using the hyper-spectral and multispectral pictures obtained from earth orbiting satellites or aerial vehicles, which can offer the variety of the fields. A geographic information system (GIS) can be used to monitor individual crops as well as crops in a given area. A management strategy known as precision farming, which combines cutting-edge information, communication, and data analysis tools into the decision-making process to increase crop yield while lowering water and nutrient losses and enhancing nutrient use efficiency, is also where remote sensing got its start. A brief overview of the use of remote sensing in sustainable agriculture development and resource conservation initiatives has been discussed in the following section.

Basic Aspects: In the initial phases of satellite remote sensing, the majority of academics concentrate on classifying land cover types using data, with crop kinds being a primary area of interest for those interested in agricultural applications. The study of agricultural remote sensing has centred more on the biophysical characteristics of plants in the past few years. Agriculture-related monitoring and analysis has long made use of remote sensing. Agricultural canopies that have been remotely sensed have yielded insightful data on a range of agronomic characteristics. Remote sensing has the advantage of being able to collect data repeatedly without damaging crop samples, which can be useful for giving important information for precision agricultural applications. The primary applications of satellite remote sensing in India are crop acreage estimation and agricultural crop output estimation. Based on the biophysical characteristics of crops and/or soils, remote sensing technology has the potential to completely change how agricultural productivity is detected and characterized. Remote sensing satellite data can be utilized for yield estimation, crop phenological information, and disturbance and stress condition identification.

Few of the applications of remote sensing in agriculture are discussed below:

Soil quality assessment by Remote Sensing

In order to sustain ecosystems and human civilization, soil is essential. In addition to being a valuable and non-renewable resource, it varies greatly in its qualities and behavior throughout the world. An inventory of the various types of soils and their distributional extent is prepared with the aid of soil surveys, allowing for the prediction of their potentialities and features. This kind of information is necessary to improve soil utilization in a sustainable manner. Conventional methods for soil surveying are expensive, labor-intensive, and time-consuming. Late developments in, information and computer technology modern technology has given forth new soil techniques. Mapping of resources such are GPS, GIS, and RS approaches can be used to analyze the various characteristics of soils across time and space. RS techniques provide useful instruments for determining and keeping track of soil parameters.

Based on the distinct spectral signatures of each soil type, spectral reflectance analysis offers insights on the properties of the soil. Estimates of soil moisture content, compaction levels, and fertility fluctuations can be obtained using thermal infrared imaging. The evaluation of soil texture, surface roughness, and moisture content is made possible using radar remote sensing. These methods enhance our knowledge of soil dynamics and help us make wise decisions in environmental studies, land management, and agriculture.

Fig 1: Some of the applications of remote sensing and GIS in Agriculture.

Depending on the light source being used, RS systems can be classified as either passive or active. Passive RS uses visible, near-infrared (VNIR), shortwave infrared (SWIR), thermal infrared (TIR), and microwave regions of the electromagnetic spectrum. Examples of these devices are imaging spectrometers and multispectral equipment. However, the majority of active sensors make use of the microwave component. For passive remote sensing, there are four different kinds of sensors. These include optical imaging spectroscopy sensors (which retrieve soil properties like organic matter, Fe-oxides, and mineralogical composition), optical TIR sensors (which estimate soil temperature), and passive microwave sensors (which estimate soil moisture). Optical multispectral sensors are used for land use and mineralogical studies. Among the active remote sensors are synthetic aperture radar (SAR) sensors, which measure the texture, moisture content, and salinity assessments), sensors for radar scatter meters (estimation of soil moisture) and LiDAR sensors (Analysis of the landscape). For soil identification there are three methods used.

Remote Sensing Method for Soil Identification		
Spectral reflectance analysis	Thermal infrared analysis	Radar remote sensing

1. **Spectral reflectance analysis:** The interactions between electromagnetic energy and the bonds in the molecules that make up the soil are responsible for the reflectance of the soil in the visible near infrared (VNIR) band (350-2500 nm). The characteristics of the soil, such as the amount of organic matter, clay minerals, iron (Fe) oxide, soluble salts, roughness, and moisture content, all have an impact on the soil's spectrum reflectance. Thus, it offers a chance to analyse soil reflectance data in multiple wavelength regions to obtain information about various soil qualities.
2. **Thermal Infrared Analysis:** With the use of non-contact, non-destructive infrared thermal imaging technology, a precise, real-time temperature distribution field of the target surface can be obtained. An objective's

surface emits invisible infrared energy, which the camera transforms into a two-dimensional visual image. This technology has been widely applied to evaluate several activities that take place at the soil surface, including evaluations of soil water repellency, preferred infiltration, evaporation, and crust formation. Variations in soil fertility and compaction can also be detected using thermal infrared imaging. Because compacted soils are less porous, there is less water infiltration and a rise in surface temperature. Thermal pattern analysis can be used to locate regions of compacted soil and evaluate how they affect plant growth.

3. **Radar Remote sensing:** Radar remote sensing extracts physical and dielectric behaviour from ground targets using electromagnetic energy backscattered from them. Radar imaging's capacity to provide all-weather and all-hour imaging is one of its main advantages. Microwave signals are used in radar remote sensing to evaluate the characteristics of soil. Because microwaves may pass through soil surfaces, it is possible to measure properties that are below the surface. Radar sensors can provide information on soil texture and moisture content by using electromagnetic waves to travel through the soil. Radar sensors use microwave signals that interact differently with different soil qualities to offer information about the texture, surface roughness, and moisture content of the soil.

Water Resource Management

Sustainable water management depends on accurate accounting of agricultural water consumption. Though there is little irrigation metering and growing strain on surface and groundwater supplies in many agricultural locations across the world, most agricultural water use remains unmonitored. Satellite remote sensing has been presented as a low-cost, scalable way around the technical, social, and political obstacles that have so far limited in situ metering to close large gaps in the monitoring of irrigation water consumption in both developed and developing nations. The region of the electromagnetic spectrum between 3.0 and 14 μm is known as thermal infrared. The fundamental idea behind thermal infrared is that anything hotter than zero degrees Celsius (K) releases electromagnetic radiation. As a result, vegetation that is warmer than absolute zero could potentially release thermal infrared, which can be detected by radiation-sensitive sensors. The emissivity of the plant is then correlated with the measured temperature, also known as radiant temperature, to determine the kinetic, or actual, temperature of the vegetation. The crop water stress index (CWSI), a measurement of the crop water status, is then calculated using this temperature, the air temperature, and the air vapor pressure. The use of thermal infrared remote sensing for irrigation scheduling has advanced significantly from the days of handheld infrared The available sensors back then were thermometers. Satellites are equipped with sensors. The primary strategy is to acquire a crop-based dataset collected remotely and then relate it to gather weather-related data and create an index it should provide an indication of the water quality of the crop. While other ground-based remote sensing techniques, such as infrared thermometers, are still used, satellite remote sensing is currently the preferred method. There are sensors on-board satellites, such as NOAA-GOES, AVHRR, Landsat TM 4 and 5, ASTER, that can provide thermal infrared pictures that can be used for irrigation scheduling.

Crop Condition Assessment

Early in the crop growth season is a critical time to gather information on crop conditions. At times, it can even hold greater significance than obtaining the precise yield after harvesting. As remote sensing applications have advanced, satellite data has emerged as the most reliable source of information for tracking the condition of large-scale crops. More than a dozen techniques based on data from remote sensing have been created over the past 20 years to monitor crop conditions in various nations, and there has been a significant improvement in the accuracy or efficacy of these techniques. Many indicators, including the normalized difference vegetation index (NDVI), vegetation condition index (VCI), and temperature condition index (TCI), have been developed using data from remote sensing. These indices are frequently used to monitor excessive soil moisture, identify droughts, analyze the effects of weather on vegetation, and gauge the productivity and health of vegetation. The NDVI statistics are widely utilized for forecasting, crop production evaluation, and vegetation monitoring. Large croplands require the use of satellite earth observation (EO) for monitoring. Due to their passive nature and reliance on sunshine, optical sensors are limited in their ability to acquire traditional satellite images. Furthermore, cloud effects provide a significant obstacle to optical imaging since light cannot pass through clouds, leading to temporal order gaps in data. On the other hand, croplands can be more effectively monitored by using radar sensors installed on satellites, which are able to both produce microwave radiation and acquire radar images of the target areas even when clouds are present. This is because the wavelengths of the microwave radiation are larger than the typical size of cloud particles, making it easy for the radiation to pass through clouds. Synthetic aperture radar (SAR) is a common type of radar system installed on satellites. Analyzing biophysical characteristics of crops such as height, LAI, biomass, and chlorophyll content is made easier with the aid of optical pictures. On the other hand, structural details, crop moisture content, and water availability are provided by radar pictures. For crop monitoring, these two sources are therefore complementary to one another. Some of the satellites which uses optical and radar images are Landsat series, Terra and Aqua, RADARSAT series, RISAT series respectively.

Because a single satellite can't cover an area continuously due to its orbit around the earth, images of a given region are only taken at intervals of a few days. Consequently, there is a temporal data gap that needs to be filled when monitoring agricultural lands. One way to close this gap is to combine images from various satellites. Now a days Sentinel Hub (SH), Google Earth Engine (GEE), Open Data Cube (ODC) etc are used for handling large scale data and for local image storage.

Pest and Disease Assessment

Pests and plant diseases pose serious risks to global agriculture and forestry. Plant protection methods must be guided by an understanding of the occurrence, distribution, and intensity of diseases and pests. Remote sensing (RS) techniques can be a crucial addition to enable the monitoring of plant diseases and pests at a coarse scale, since traditional field scouting of plant diseases and pests is labour-intensive, prone to subjectivity, and typically exhibits low efficiency. One way to think of remote sensing plant illnesses and pests is as a kind of "radiodiagnosis" for plants, which effectively provides noncontact, spatially constant monitoring of diseases and pests. A vast range of RS data have been used to identify illnesses and pests as a result of the quick advancements in computer science and sensing technologies. Research on sensing systems, feature extraction, and algorithms has been conducted at various scales over the past few decades, opening up new opportunities for the remote monitoring of plant diseases and pests. This strategy has improved our understanding of the agriculture system in many ways by strengthening the links between plant pathology ideas and RS methodologies. These RS devices can collect data from microwave to gamma rays and can operate with both passive and active radiation. In an attempt to effectively identify and track plant diseases and pests, numerous RS systems have been implemented to capture the infection symptoms (such as scabs, pustules, etc.), physiological reactions (such as variations in pigment content, water content, etc.), and structural alterations (such as canopy structure, landscape structure, etc.) brought on by plant diseases and pests (Abd El Ghany 2020).

Regarding the problem of plant disease and pest monitoring, sensing systems can be broadly divided into three groups based on sensing principles and technological maturity:

Visible and near infrared (VIS-SWIR) spectral systems; Fluorescence and thermal systems; and Synthetic aperture radar (SAR) and light Detection and Ranging equipment (Lidar) systems.

1. **Visible and near infrared (VIS-SWIR) spectral systems:** The visible and near-infrared (VIS-NIR) bands contain the majority of solar radiation, hence sensors operating in these bands are typically able to obtain high-quality data with a comparatively high signal-to-noise ratio. The VIS-NIR sensors are a well-established and reasonably priced technology that can be easily accessed by several remote sensing platforms, including aerial, satellite, and ground-based platforms. This makes them a promising option for monitoring pests and plant diseases. Different physiological changes and signs of diseases and pests (e.g., pigment degradation, cellular damage, scabs, etc.) exhibit distinct reactions in spectral reflectance in the VIS-NIR spectrum domain (Prabhakar et al 2022).
2. **Fluorescence and thermal systems:** Plant respiration and photosynthetic processes can be monitored using thermal infrared and fluorescence remote

sensing (RS) systems, enabling presymptomatic detection of pests and diseases. Laser-induced fluorescence (LIF) is the most promising method for measuring plant fluorescence response. Commercialized fluorometers, like the portable chlorophyll fluorometer PAM series (PAM2100, PAM2500, IMAGINE-PAM, etc.) from Heinz Walz GmbH, use the saturation pulse technique and induced LED light sources at fixed wavelengths (655 nm and 735 nm) to record some kinetics curves and parameters and to generate some important fluorescence indicators, such as the maximal quantum yield (Fv/Fm), nonphotochemical fluorescence quenching (NPQ), the operating efficiency of PSII etc.

Apart from the fluorescence response, diseases and pests can also alter the respiration and evapotranspiration rates of plants, resulting in discernible changes in the levels of thermal emission (Nicolas, 2004). Stoll et al. (2008) discovered that thermal photography may effectively monitor the abnormalities in temperature at an early growth stage in order to detect an attack of the fungal disease *Plasmopara viticola* in grapevines. Falkenberg et al. (2007) effectively distinguished between biotic (root rot) and abiotic (drought) conditions in cotton using thermal image analysis. Other research has also verified the efficacy of thermal and fluorescent approaches in the early detection of pests and plant diseases (e.g., Chaerle and Van Der Straeten, 2000; Murchie and Lawson, 2013; Raza et al., 2015).

3. **Synthetic aperture radar (SAR) and light Detection and Ranging equipment (Lidar) systems:** Since many plant diseases and pests thrive in cloudy and rainy regions, clouds obstruct the application of VIS-NIR RS images. As an active method of remote sensing, the cloud penetrating capability of SAR enables monitoring in all weather conditions. Currently, few studies have used SAR and Lidar RS in the monitoring of plant diseases and pests. This may be because there is a weak correlation between SAR and Lidar parameters and the symptoms of plant diseases and pests. Nevertheless, SAR and Lidar can be used to delineate plant characteristics and their environments, which may be utilized in the monitoring of diseases and pests (Singh et al., 2007). The Lidar system can gather comprehensive data on canopy morphology. Plant canopy point cloud data, gathered by Lidar sensors, allows for the reconstruction of the canopy's three-dimensional structure, revealing damages or biomass loss brought on by pests or illnesses. Certain plant diseases, such the peanut sclerotinia blight, can drastically change the canopy morphology of plants and may be picked up by Lidar sensors.

Assessing and managing plant protection requires timely, accurate, and reliable disease and pest monitoring over large regions. Many RS techniques have been developed in the last few decades to monitor pests and plant diseases. These techniques have shown a lot of promise in assisting traditional, time-consuming inspection.

Conclusion

Because agricultural areas can vary greatly in terms of geography, elevation, moisture content, and nutrient content, the empirical link between source and sink is extremely brittle. It is necessary to develop a standardized link, crop-specific, between the spectrum characteristics and physio-chemical parameters. A satellite's initial installation costs are now exceedingly costly and beyond the reach of the average person. Ground truthing is still important because using remote sensing alone for precision agriculture can be confusing and wasteful of resources. This increases the complexity and expense of using remotely sensed data collected from satellites for policy makers. The agricultural sciences can greatly benefit from remote sensing applications, which have a lot of theoretical underpinnings and potential to increase productivity, predict vulnerability, and analyze impacts in the context of rapidly changing climate conditions.

References

Abd El-Ghany, N.M., Abd El-Aziz, S.E. & Marei, S.S. 2020 A review: Application of remote sensing as a promising strategy for insect pests and diseases management. Environ Sci Pollut Res 27, 33503-33515

Chaerle. L, Dominique Van Der Straeten, 2000 Imaging techniques and the early detection of plant stress, Trends in Plant Science, 5, 11, 495-501.

Falkenberg N, Giovanni Piccinni, J. Tom Cothren, Daniel I. Leskovar, Charlie M. Rush, 2007 Remote sensing of biotic and abiotic stress for irrigation management of cotton, Agricultural Water Management, 87, (1), 23-31.

Murchie, E.H., Lawson, T., 2013 Chlorophyll fluorescence analysis: a guide to good practice and understanding some new applications, *Journal of Experimental Botany*, 64, (13), 3983–3998.

Nicolas, H 2004 Using remote sensing to determine of the date of a fungicide application on winter wheat, Crop Protection, 23, (9), 853-863.

Prabhakar, M., Thirupathi, M., Mani, M. 2022. Principles and Application of Remote sensing in crop pest management. In: Mani, M.(eds) Trends in Horticultural Entomology. 157-183.

Raza S-e-A, Prince G, Clarkson JP, Rajpoot NM (2015) Automatic Detection of Diseased Tomato Plants Using Thermal and Stereo Visible Light Images. PLoS ONE 10(4): e0123262.

Singh, D Sao, R., Singh K P 2007, A remote sensing assessment of pest infestation on sorghum, Advances in Space Research, 39, (1) 155-163

Stoll, M., Schultz, H.R., Baecker, G. et al. 2008 Early pathogen detection under different water status and the assessment of spray application in vineyards through the use of thermal imagery. Precision Agric 9, 407–417.

Zhanga J ,Huangb Y, Puc R , Pablo Gonzalez-Morenod , Yuane L, Kaihua Wua, Wenjiang Huangf ., 2019 Monitoring plant diseases and pests through remote sensing technology: A review, Computers and Electronics in Agriculture 165, 2019, 104943

CHAPTER 20

DIVERSITY OF WEED FLORA IN KHAIRAGARH DISTRICT OF CHHATISGARH

Manisha Nayak[1] & Bhabita Mandavi[1]

[1]Department of Chemistry & Botany, Rani Rashmi Devi Singh Govt. College, Khairagarh C.G,.manishanayak1994@gmail.com

Abstract

Khairagarh which is a music city came into existence in 03 September 2022. For the first time after coming into existence study of diversity of weed flora has been done. For this, field study was conducted during the Rabi and Kharif has been done in which sample were collected from weed flora in rice field and plains of Khairagarh. During the survey, we found a total of 98 plant species. We identified and documented them. We found 80 genera and 98 species belonging to 27 families. The most dominant family was Poaceae 17 species each and co-dominant family was Asteraceae 13 species, followed by Fabaceae 12 species, Amaranthaceae 9 species, Convolvulaceae and solanaceae 6 species each, Lamiaceae 5 species each, Cyperaceae 4 species each, Euphorbiaceae and Tiliaceae 3 species each, Malvaceae and Rubiaceae 2 species each, and the remaining Acanthaceae, Amaryllidaceae, Apiaceae, Asclepiadaceae, Brassicaceae, Cannabaceae, Capparidaceae, Commelinaceae, Gentianaceae, Martyniaceae, Oxalidaceae, papaveraceae, Portulaceae, Sapindaceae, Verbanaceae 1 species each respectively. In the present study, we also found some plants that play an important role in biodiversity conservation. They are as follows: *Achyranthes aspera, Cassia* species, *Euphorbia* species and *Lantana camara* are used in the preparation of vermicompost. The weedy grass Panicum species, which fixes higher nitrogen, and *Echinochloa* species, *Cynodon dactylon,* prevent soil erosion and *Arundo donax* is used in paper making. Diversity of weeds talks about the large variety of weeds in nature, and since weeds have so many uses, it is very important to conserve their diversity.

Keywords: Soil, Weed flora, Diversity, Khairagarh, C.G.

Introduction

Earth is only a planet on which life exists; it is known as the "blue of green" planet since weather covers 71% of its surface and the remaining 29% is made up of land. Simply said, the environment is the space where humans and other living things exist (Singh & Upadhyay 2018). The organic and inorganic components that make up the Earth's surface and act as a growing substrate for plants are referred to as soil (Dewangan et.al.,2022). We refer to Chhattisgarh as the "rice bowl". The state of Chhattisgarh is located between latitudes 17 46′–24 8′ and longitudes 80 15′–84 24′ E. The state's overall area, or 1,35,194 sq. km, is 136034.28 km2 from north to south and 336 km from east to west. The nutritional status of soil is based on the availability of both macro and micronutrients (Singh et.al.,2018). The temperature, moisture content, microbiological activity, and nutrient absorption are all influenced by the physical and chemical characteristics of the soil. The soil's macro- and micronutrients contribute to the forage crop's qualitative and quantitative output. Therefore, it's crucial to keep an eye on the condition of the soil and determine which farming system is best sustainable for the region.

Three sorts of plants exist based on the fertility of the soil. i) wild plants, which develop naturally outside of human-damaged habitats; ii) weeds, which flourish in human-damaged habitats; and iii) domesticates, or crops, which are artificially reproduced and frequently need human care and cultivation to grow and utilize environmental resources (Tiwari et.al.,2016). Professor Alden Springer Crafts, a renowned weed expert, stated that "in the beginning there were no weeds" in one of his earliest official publications on plant ecology. In other words, he was saying that the classification of some plants as weeds was merely an anthropomorphic practice that resulted from human judgment of resources according to their apparent usefulness or futility (Nayak et.al.,2023). Plants that grow in unwanted areas are known as weeds. Thus, weeds imply that they are unattractive, unwelcome plants that are potentially toxic and destructive, as well as frequently abundant and persistent. These plants harm crop plants, obstruct agricultural operations, require more labor, raise cultivation costs, degrade the quality of agricultural output, and ultimately lower crop yields. One of the most significant contributing factors to low rice output is the loss resulting from weed infestation (Shrivastava et.al., 2014). Most dangerous biological threat to agricultural output is weeds, which account for 33% of pest-related losses. Due to their tremendous adaptability, weeds can grow more quickly, take over the crop habitat, and lower crop yields (Pradhan et.al.,2022). mostly seen that every plant has a monetary worth as food, decoration, or therapeutics. In addition, weeds are crucial for a number of other reasons, including soil health, ecosystem balance, economic effect, insect and disease control, and agricultural production.

Material and Method

Study Area

The current study was conducted in Chhattisgarh's Khairagarh-Chuikhadan-Gandai district. One of the districts in the west of the state of Chhattisgarh is Khairagarh. The district was established on September 3, 2022. Rajnandgaon was the district of its father. This location's latitude and longitude are 21.4183° N and 80.97 ° E, respectively. Khairagarh has a 1033 km2 area. Because of Indira Kala Sangeet University, Asia's first music university, is located in Khairagarh, the city is referred to as the "**City of Music.**" The districts of Kabirdham, Rajnandgaon, Bemetara, and Durg encircle Khairagarh. In Khairagarh, there is about 900 millimeters of rainfall on average year.

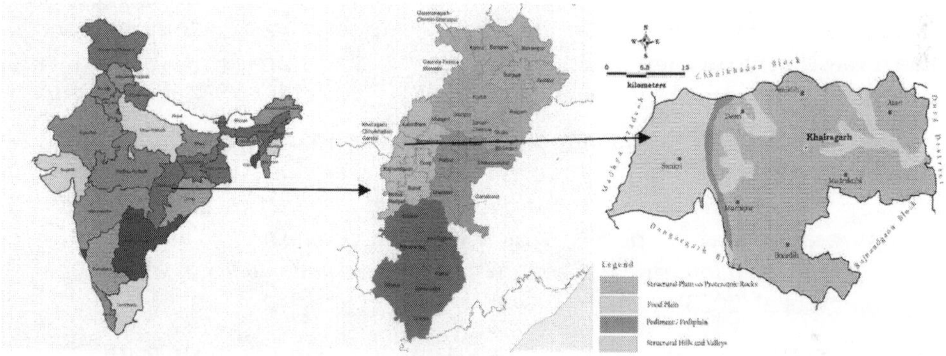

Fig 1: The Geographical map of Khairagarh district, Chhattisgarh

Soil and Climate

Depending on the geology, height, slope, climate, vegetation, and soil chemistry and biology of a given site, there are many differences in soil texture, color, and character. Except for a few farmed areas, the soil is typically deep clay loam, high in potassium, mildly acidic, and has a medium phosphorus and nitrogen concentration. This region experiences a scorching summer from mid-April to mid-June. Apart from these two months, the weather is always nice. The average annual rainfall in Khairagarh is approximately 900 mm.

Field Survey

The current study includes a thorough field survey conducted during the study period to identify plants. Many months of the Rabi and Kharif agricultural seasons, as well as the wild species of the district, were the subjects of extensive field surveys in 2023 July–July 2024. The collection and documentation of specimens were done using random sampling techniques. This survey's primary goal was to gather data on the different kinds of weed flora that can be found there. In addition, we took pictures and gathered specimens of various weed species all across Khairagarh. In light of the "Flora of British India" (Hooker, 1872–1897) as well as additional state, regional, and local floras, a provisional identification was produced. Bentham and Hooker's taxonomy (1862–83) was followed by all plant families, with a few exclusions made to account for more recent changes made in response to Cronquist (1968).

Results and Discussion

During the survey, we found a total of 98 plant species. We identified and documented them. We found 80 genera and 98 species belonging to 27 families. The most dominant family was Poaceae 17 species each and co-dominant family was Asteraceae 13 species followed by Fabaceae 12 species, Amaranthaceae 9 species, Convolvulaceae and Solanaceae 6 species each, Lamiaceae 5 species each, Cyperaceae 4 species each, Euphorbiaceae and Tiliaceae 3 species each, Malvaceae and Rubiaceae 2 species each, and the remaining Acanthaceae, Amaryllidaceae, Apiaceae, Asclepiadaceae, Brassicaceae, Cannabaceae, Capparidaceae, Commelinaceae, Gentianaceae, Martyniaceae, Oxalidaceae, Papaveraceae, Portulaceae, Sapindaceae, Verbanaceae 1 species each respectievely.(Table:1; Table :2;)

In the present study, we also found some plants that play an important role in biodiversity conservation. They are as follows: *Achyranthes aspera*, *Cassia species*, *Euphorbia* species and *Lantana camara* are used in the preparation of vermicompost. The weedy grass *Panicum* species, which fixes higher nitrogen, and *Echinochloa* species, *Cynodon dactylon*, prevent soil erosion and *Arundo donax* is used in paper making (Singh et.al.,2023).

Table 1: Diversity of weed flora in Khairagarh district of Chhattisgarh.

S.No.	Common name	Botanical name	Family
1.	Devils housewhip	*Achyranthes Aspera* Linn.	Amaranthaceae
2.	Desert cotton	*Aerva javanica* (Burm.f.) Shult.	Amaranthaceae
3.	Goat weed	*Ageratum conyzoides* L.	Asteraceae
4.	Red calico plant	*Alternanthera bettzickiana* (Regel) G. Nicholson	Amaranthaceae
5.	Alligator weed	*Alternanthera philoxeroides* (Mart.) Griseb	Amaranthaceae
6.	Sessile joyweed	*Alternanthera sessilis* (L.) R.Br. *exDC.*	Amaranthaceae
7.	Alyce clover	*Alysicarpus vaginalis* (L.) DC.	Fabaceae
8.	Spiny amaranth	*Amaranthus spinosus L.*	Amaranthaceae
9.	Slender amaranth	*Amaranthus viridis* L.	Amaranthaceae
10.	Prickly poppy	*Argemone Mexicanav* L. Mexican	Papaveraceae
11.	Elephant grass	*Arundo donax* L.	Poaceae
12.	Vajradanti	*Barleria prionitis*	Acanthaceae
13.	Bathuva	*Chenopodium album Linn.*	Amaranthaceae
14.	Jangli Muli	*Blumea lacera* L.	Asteraceae
15.	Sheda grass	*Bothriochloa pertusa* (L.) A. Camus.	Poaceae

S.No.	Common name	Botanical name	Family
16.	Swallowwort	*Calotropis procera* (Willd.) R. Br	Asclepiadaceae
17.	Hemp	*Cannabis sativa* L.	Cannabaceae
18.	Bhui neem	*Canscora diffusa* (Vahl) R.Br. ex Roem. & Schult.	Gentianaceae
19.	Balloon vine	*Cardiospermum halicacabum* Linn	Sapindaceae
20.	Dwarf cassia	*Cassia chamaecrista* L. var	Fabaceae
21.	Pigs senna	*Cassia occidentalis* L	Fabaceae
22.	Buffel grass	*Cenchrus ciliaris* Linn	Poaceae
23.	Pennywort Asiatic	*Centella asiatica* (L.) Urban	Apiaceae
24.	Butterfly pea	*Centrosema pubescens* Benth	Fabaceae
25.	Purple chloris	*Chloris barbata* Sw	Poaceae
26.	Siam weed	*Chromolaena odorata* (L.) R.M. King	Asteraceae
27.	Canada thistle	*Cirsium arvense* (L.) Scop	Asteraceae
28.	Tick weed	*Cleome viscosa* L	Capparidaceae
29.	Dayflower	*Commelina benghalensis* Linn	Commelinaceae
30.	Field bindweed	*Convolvulus arvensis* Linn.	Convolvulaceae
31.	East Indian Jew mallow	*Corchorus aestuans* L	Tiliaceae
32.	White jute	*Corchorus capsularis* L	Tiliaceae
33.	Tall wild jute	*Corchorus fascicularis* Lam.	Tiliaceae
34.	Lesserswine cress	*Coronopus didymus* (L.) Sm.	Brassicaceae
35.	Indian hemp	*Crotalari juncea* Linn.	Fabaceae
36.	Jungke tulsi	*Croton bonplandianum* Bail	Euphorbiaceae
37.	Dubgrass	*Cynodon dactylon* Linn	Poaceae
38.	Bunchy sedge	*Cyperus polystachyos* Rottb	Cyperaceae
39.	Nagramotha	*Cyperus rotundus* Linn.	Cyperaceae
40.	Cow poot grass	*Dactyloctenium aegyptium* (L.) Willd.)	Poaceae
41.	Thorn apple	*Datura metel* Linn.	Solanaceae
42.	Klebergs bluestem	*Dichanthium annulatum* (Forssk.) Stapf)	Poaceae
43.	Jungle rice	*Echinochloa colonum* (L.)	Poaceae
44.	Barnyrd grass	*Echinochloa crus-galli* (L.) P. Beauv	Poaceae
45.	Indian globethistle	*Echinops echinatus* Roxb.	Asteraceae
46.	False daisy	*Eclipta alba* (L.)	Asteraceae
47.	Goose grass	*Eleusine indica* (L.) Gaertn.	Poaceae
48.	Red tassel flower	*Emilia sonchifolia* (L.) DC. Ex DC	Asteraceae
49.	Japanese lovegrass	*Eragrostis tenella* (L.) Beauv. ex R.&S.	Poaceae
50.	Dudhi	*Euphorbia hirta* L.	Euphorbiaceae

S.No.	Common name	Botanical name	Family
51.	Dwarf Morning Glory	*Evolvulus alsinoides* L.	Convolvulaceae
52.	Bindweed	*Evolvulus nummularius* (L.)	Convolvulaceae
53.	Forked fimbry	*Fimbristylis dichotoma* (L.) Vahl.	Cyperaceae
54.	Grass like fimbry	*Fimbristylis miliacea* (L.) Vahl	Cyperaceae
55.	Gul-e-makhmal	*Gomphrena celosioides* Mart	Amaranthaceae
56.	Ban tulsi	*Hyptis suaveolens* (L.) Poit	Lamiaceae
57.	Cogon grass	*Imperata cylindrica* (L.)	Poaceae
58.	Heart leaf indigo	*Indigofera cordifolia.* B. Heyne ex Roth	Fabaceae
59.	Water sinach	*Ipomoea aquatic* forssk	Convolvulaceae
60.	Red Morning Glory	*Ipomoea coccinea* Linn.	Convolvulaceae
61.	Lilac bell	*Ipomoea turbinate* Lag	Convolvulaceae
62.	Indian muraina grass	*Ischaemum indicum* (Houtt.) Merr.	Poaceae
63.	Beach launaea	*Lactuca sarmentosa* (Willd.) DC	Asteraceae
64.	Wild sage	*Lantana camara* L.	Verbanaceae
65.	Yellow pea	*Lathyrus aphaca* L.	Fabaceae
66.	Lions ear	*Leonotis nepetaefolia* (L.) R. Br	Lamiaceae
67.	Common leucas	*Leucas aspera* (Willd.) Linn	Lamiaceae
68.	White wort	*Leucas martinicensis* (Jacq.)	Lamiaceae
69.	Ceylon slitworp	*Leucas zeylanica* (L.) W.T. Aiton	Lamiaceae
70.	Tiger's claw	*Martynia annua* L	Martyniaceae
71.	Rough medik	*Medicago denticulate* willd	Fabaceae
72.	Touch me not	*Mimosa pudica* L	Fabaceae
73.	Tropical girdlepod	*Mitracarpus villosus (Sw.)* DC.	Rubiaceae
74.	Amrul	*Oxalis corniculata* L	Oxalidaceae
75.	Blue Panic grass	*Panicum antidotale* Retz.	Poaceae
76.	Gajar grass	*Parthenium hysterophorus* Linn.	Asteraceae
77.	Stonebreaker	*Phyllanthus niruri* L.	Euphorbiaceae
78.	Rashary	*Physalis minima Linn.*	Solanaceae
79.	Cape goose berry	*Physalis peruviana* L	Solanaceae
80.	Annual Bluegrass	*Poa annua* Linn.	Poaceae
81.	Annual rabbits foot grass	*Polypogon fugax* f. muticus (I.C. Chung)	Poaceae
82.	Common purslane	*Portulaca oleracea* Linn.	Portulaceae
83.	Babchi	*Psoralea corylifolia* Linn	Fabaceae
84.	Sicklepod	*Senna obtusifolia* (L.) H.S. Irwin & *Barneby*	Fabaceae

S.No.	Common name	Botanical name	Family
85.	Green foxtail	*Setaria viridis* (L.)	Poaceae
86.	Wireweed	*Sida acuta burm* (F.)	Malvaceae
87.	Country mallow	*Sida cordifolia* Linn	Malvaceae
88.	Lam Sticky night shad	*Solanum sisymbriifolium*	Solanaceae
89.	Tropical soda	*Solanum viarum* dunal	Solanaceae
90.	Yellow berried right Shade	*Solanum xanthocarpum* schrad	Solanaceae
91.	Smooth aqoethistle	*Sonchus oleraceus* (L.)	Asteraceae
92.	Tiny false button weed	*Spermacoce pusilla* Wall.	Rubiaceae
93.	Gorukhmundi	*Sphaeranthus indicus* Linn	Asteraceae
94.	Paracress	*Spilanthes calva* DC	Asteraceae
95.	Tephrosia	*Tephrosia purpurea* (L.) Pers.	Fabaceae
96.	Tridax daisy	*Tridax procumbens* Linn.	Asteraceae
97.	Sleepy morning	*Waltheria indica* L.	Malvaceae
98.	Rosy rain lily	*Zephyranrhus rosea* Lind.	Amaryllidaceae

Table 2: List of Distribution of Family

S.No.	Family	No. of genus	No. of species
1.	Acanthaceae	1	1
2.	Amaranthaceae	6	9
3.	Amaryllidaceae	1	1
4.	Apiaceae	1	1
5.	Asclepiadaceae	1	1
6.	Asteraceae	13	13
7.	Brassicaceae	1	1
8.	Cannabaceae	1	1
9.	Capparidaceae	1	1
10.	Commelinaceae	1	1
11.	Convolvulaceae	3	6
12.	Cyperaceae	2	4
13.	Euphorbiaceae	3	3
14.	Fabaceae	11	12
15.	Gentianaceae	1	1
16.	Lamiaceae	3	5
17.	Malvaceae	2	3
18.	Martyniaceae	1	1

S.No.	Family	No. of genus	No. of species
19.	Oxalidaceae	1	1
20.	Papaveraceae	1	1
21.	Poaceae	16	17
22.	Portulaceae	1	1
23.	Rubiaceae	2	2
24.	Sapindaceae	1	1
25.	Solanaceae	3	6
26.	Tiliaceae	1	3
27.	Verbanaceae	1	1

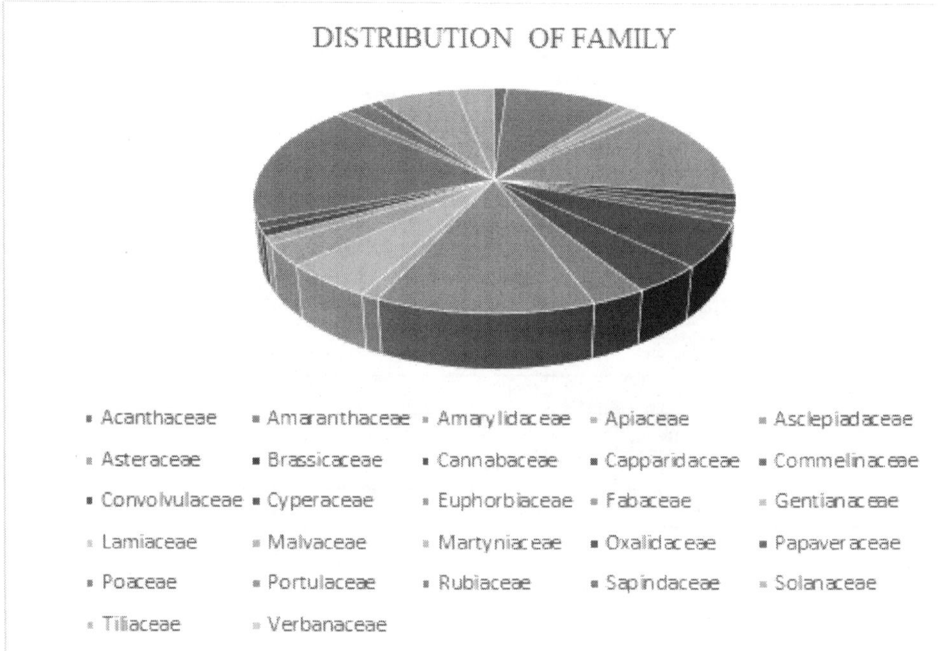

Fig: 2 Distribution of family

Table 3: Serial no. of plant species belonging to different families

S.No.	Family	Serial no. of genus
1.	Acanthaceae	12
2.	Amaranthaceae	1,2,4,5,6,8,9,13,55
3.	Amaryllidaceae	98
4.	Apiaceae	23
5.	Asclepiadaceae	16
6.	Asteraceae	3,14,26,27,45,46,48,63,76,91,93,94,96

S.No.	Family	Serial no. of genus
7.	Brassicaceae	34
8.	Cannabaceae	17
9.	Capparidaceae	28
10.	Commelinaceae	29
11.	Convolvulaceae	30,51,52,59,60,61
12.	Cyperaceae	38,39,53,54
13.	Euphorbiaceae	36,50,77
14.	Fabaceae	7,20,21,24,35,58,65,71,72,83,84,95
15.	Gentianaceae	18
16.	Lamiaceae	56,66,67,68,69
17.	Malvaceae	86,87,97
18.	Martyniaceae	70
19.	Oxalidaceae	74
20.	Papaveraceae	10
21.	Poaceae	11,15,22,25,37,40,42,43,44,47,49,57,62,75,80,81,85
22.	Portulaceae	82
23.	Rubiaceae	73,92
24.	Sapindaceae	19
25.	Solanaceae	41,78,79,88,89,90
26.	Tiliaceae	31,32,33
27.	Verbanaceae	64

Conclusion

In the natural world, everything is valuable since nothing is wasted. The weeds are growing effectively since they are the most sustainable and flexible plants. Sustainable product conserve both soil and biodiversity. When research was done on weeds, it was found that weeds are very useful and can be used as a source of macro and micronutrients of plants, heavy metal detection, preventing soil erosion and other work like biodiversity conservation control of insect pests etc. Diversity of weeds talks about the large variety of weeds in nature, and since weeds have so many uses, it is very important to conserve their diversity.

References

Bisht, A. S. (2017). Weed floral diversity of medicinal value in terraces of horticulture crop fields in Bharsar, Uttarakhand, India. Indian Journal of Plant Genetic Resources, 30(02), 153-161

Dewangan, S. K., Kehri, D. Preeti. & Yadav, A. (2022). Study The Physico-Chemical Properties of Brown Soil of Gaura Village of Surajpur District, Surguja Division of Chhattisgarh, India. International Journal of Engineering Inventions, 11(11), 80-83

Hassan, S. (2020). Positive aspects of weeds as herbal remedies and medicinal plants. Journal of Research in Weed Science, 3(1), pp 57-70

Krishna, B., Sahu, K. K., Tedia, K., Singh, C. B., & Kumar, R. (2018). Evaluation of Soil Fertility Status of Palari Block under Baloda Bazar District of Chhattisgarh. Journal of Pharmacognosy and Phytochemistry, 7(2S), 243-246

Kumar, A., Mishra, V. N., Srivastav, L. K., & Banwasi, R. (2014). Evaluations of soil fertility status of available major nutrients (N, P & K) and micro nutrients (Fe, Mn, Cu & Zn) in Vertisol of Kabeerdham District of Chhattisgarh, India. International Journal of Interdisciplinary and Multidisciplinary Studies, 1(10), 72-79

Kumar, U., Mishra, V. N., & Kumar, N. (2018). Comparative study of soil quality indicators for rice-based cropping system in Vertisols of central plain of Chhattisgarh. Asian Journal of Soil Science, 13(2), 136-147

Lal, S. (2021). Diversity of weed flora of Dongargaon, Dist Rajnandgaon (Chhattisgarh), India. International Journal of Research and Analytical Reviews (IJRAR), 8(3), 783-793

Nayak, S., Biswas, D., & Abhismita Roy, D. V. R. (2023). Study of Weed Flora in the Central Region of Chhattisgarh: A Scientific Survey. Bull. Env. Pharmacol. Life Sci, 12, 47-53

Sharma, M., & Sharma, R. P. (2017). Diversity of edible wild plants of Pendra road forest region of Chhattisgarh. International Journal of Advanced Educational Research, 2(4), 24-28

Shrivastava, A. K., Tikariha, A., & Patra, S. (2014). Seasonal and floristic biodiversity of weeds growing in Chunkatta and Bhilai area of Chhattisgarh, India. Internantional Journal of Current Microbiology and Applied Science. 3(6), 318-326

Singh, R., Upadhyay, M. (2018). Analytical Study of Soil Quality of NCPH Colliery of Koriya District, SSRG International Journal of Applied Chemistry, 5(2), 7-10

Singh, T., Choudhary, A., & Kaur, S. (2023). Weeds can help in biodiversity and soil conservation .Indian journal of Weed Science, (2023) 55(2): 133-140

Soni MY, Sharma RP. Survey of Weed Flora in Rice Fields of Bilaspur District, Chhattisgarh. J Plant Sci Res. 2022;9(2): 220

Sreena, K. S., Yadahalli, G. S., Chittapur, B. M., Ajayakumar, M. Y., Bhat, S. N., & Naik, N. (2023). Significance of Weed Flora Identification and its Management on Bt Cotton (Gossypium hirsutum L.) Growing Tract of Tungabhadra Command Area. International Journal of Plant & Soil Science, 35(18), 781-789

Tiwari, D., Upadhyay, S., & Paliwal, S. J. J. A. V. S. (2016). Diversity of weed flora of Bharsar, Pauri Garhwal (Uttarakhand), India. J. Agric. Veter. Sci, 9, 1-9

APPENDIX

Photoplate: Some Picture Captured of Weed Flora

Amaranthus viridis L.

Commelina benghalensis Linn

Cynodon dactylon Linn

Datura metel Linn.

Echinochloa crus-galli (L.) P. Beauv

Euphorbia hirta L.

Ipomoea aquatic forssk

Lantana camara L.

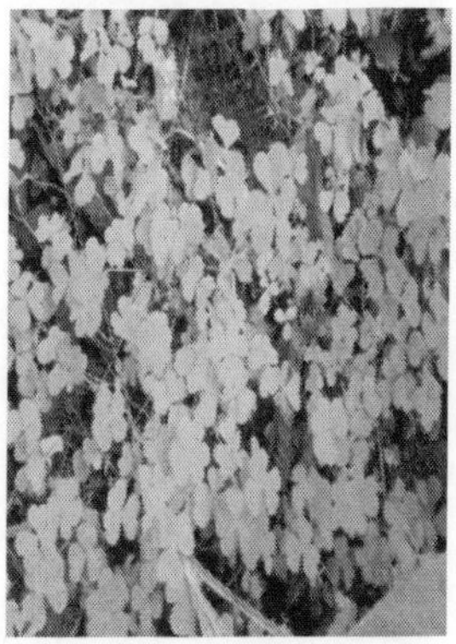

Oxalis corniculata L

Phyllanthus niruri L.

Senna obtusifolia (L.) H.S. Irwin & Barneby

Setaria viridis (L.)

Sida cordifolia Linn

Tephrosia purpurea (L.) Pers.

Tridax procumbens Linn.

Zephyranrhus rosea Lind.

CHAPTER 21

THE ENVIRONMENTAL IMPACT OF PARASITIC PLANTS: A GROWING CHALLENGE

Vandana Mishra1, Dr. Sweta Gaikwad2

[1]*Research Scholar, Shri Shankaracharya Professional University, Bhilai, Chhattisgarh, India*

[2]*Asst. Professor, Shri Shankaracharya Professional University, Bhilai, Chhattisgarh, India*

swetagaikwad7@gmail.com

Abstract

Plant parasites are common infections that affect a wide variety of plant species and result in severe losses to agriculture. By uniting their tissues and tying their vasculature to that of their hosts, they effectively extract carbohydrates, water, and nutrients from their hosts.Parasitic plants, often overlooked in ecological studies, represent a significant and growing challenge to ecosystems worldwide. These unique organisms, which rely on host plants for water, nutrients, and sometimes photosynthesis, have far-reaching impacts on biodiversity, agriculture, and natural habitats. This chapter search through the environmental implications of parasitic plants, exploring their role in disrupting plant communities, altering species compositions, and affecting the overall health of ecosystems. The chapter also examines the economic consequences of parasitic infestations on agricultural productivity, particularly in regions where food security is already insecure. Furthermore, it discusses the complex interactions between parasitic plants and climate change. Through case studies and recent research, this chapter underscores the need for integrated management strategies and increased awareness of the ecological and economic threats posed by parasitic plants. The discussion ultimately calls for a deeper understanding of parasitic plants' role in global environmental change and emphasizes the importance of sustainable approaches to mitigating their impact.

Key words: Parasitic plants, host-parasite relation, biodiversity, ecological impact, agriculture, economic threat.

Introduction

Parasitic plants get all or most of their nourishment from another plant, known as the host, without benefiting the host or, in certain situations, seriously harming it. The haustorium, a specialized organ that enters the host and creates a vascular union between the plants, is the distinguishing structural characteristic of a parasitic plant. Parasitic plants comprise approximately 4500 species which accounts for 1% of angiosperms (Heide-Jørgensen, 2008; Nickrent, 2012). They are different from others like lianas, climbing vines, aerophytes, and epiphytes (Westwood, 2024). Aerophytes are supported by other plants, but they are not parasitic because they use other plants only as a support system, not as a source of nutrients or water. The mycoheterotrophs are another category of plants that are occasionally mistaken for parasites. Mycoheterotrophs may not have chlorophyll or the ability to photosynthesize, much like parasitic plants, yet they coexist symbiotically with fungi that eat decomposing vegetation or autotrophic (self-feeding) plants. Due to their absence of haustoria and apparent lack of harm to the fungi, these plants are not considered parasitic.

These plants are a taxonomically diverse group of angiosperms that rely partially or completely on host plants for carbon, nutrients and water. They do this by attaching themselves to the roots or shoots of their hosts using haustoria, which are specialized structures that allow them to penetrate xylem and/or form close connections with phloem (Press & phoenix, 2005). The performance of hosts is frequently drastically reduced by parasitism, which alters the competitive relationships between host and non-host plants and has an impact on the structure, variety, vegetation cycling, and zonation of communities (Pennings & Callaway, 1996). Families of angiosperms that are parasitic include Rafflesiaceae, Balanophoraceae, and Orobanchaceae. Despite being suggested to be parasitic, one gymnosperm species, *Parasitaxus susta*, may really be a mycoheterotroph because it seems to involve a fungal symbiont.

Such plants can have a significant impact not only on community plants but also on individual plants. Numerous other species, such as mycorrhizal fungi, birds, insect herbivores, and other parasites, may also be impacted, either directly or indirectly.Because many parasitic plants can have both top-down and bottom-up effects, there is a wide spectrum of affects that can occur. For instance, when the host is a common resource that may be limited, parasitic plants may face competition from other consumers (Pennings & Callaway, 2002). Fungi and insects can exploit host plants that are damaged by parasite plants; herbivorous insects and mammals eat the foliage of parasitic plants; and birds that hunt for food consume the berries of mistletoe (Parker & Riches, 1993; Aukema, 2003). Parasitic plants significantly impact ecological systems by weakening or killing their host plants, leading to reduced plant diversity and altered community structures. By extracting water and nutrients, they can suppress dominant species, creating opportunities for less competitive plants and potentially altering the entire ecosystem's dynamics. Climate change may cause parasite ranges to shift along with their host species, and there is evidence that some tree parasitic plant species have already seen an extension in their altitudinal range due to warmer temperatures (Dobbertin et al., 2005). These changes can have cascading effects, influencing not only plant communities but also the animals that depend on them, thereby disrupting food webs and habitat structures. Additionally, such plants can affect nutrient cycling and primary productivity, leading to broader implications for ecosystem function and resilience. As climate change progresses, the influence of parasitic plants are expected to intensify, posing further challenges to ecological stability. In light of the profound and often underappreciated impacts of these plants on both natural and agricultural systems, it is crucial to enhance our understanding of these species and their interactions within ecosystems.

More research on the role of these parasites in forest processes is required to understand how we should manage these infections in the future. Bell and Adams (2011) provide a thorough overview of our current knowledge of the effects of parasitic plants on tree physiology and the few studies of larger-scale forest implications. As global environmental changes continue to reshape landscapes, the influence of parasitic plants is likely to grow, making it imperative to develop effective management strategies. This chapter aims to explore these challenges, providing a comprehensive overview of the environmental consequences of parasitic plants and offering insights into potential solutions. By deepening our knowledge and taking proactive steps, we can mitigate the threats posed by these complex organisms and protect the integrity of our ecosystems.

Biology of Parasitic Plants

These unique group of plants, unlike most other plants tap into the vascular system of host plants, usually the xylem (water-conducting tissue) or phloem (sugar-conducting tissue), to extract water, nutrients, and sometimes even photosynthates.

Key Characteristics

- **Haustoria**: Specialized structures that penetrate the host plant's tissues and connect to its vascular system.
- **Hemi and Holoparasites**: Some parasitic plants are partially autotrophic (hemiparasites), meaning they can photosynthesize but still rely on a host for water and nutrients. Others are fully heterotrophic (holoparasites) and lack chlorophyll, relying entirely on their host for all nutrients.
- **Host Specificity**: They vary in their host specificity, some being generalists, capable of parasitizing many species, and others being specialists, restricted to a single or a few closely related host species.

Examples

- **Mistletoe *(Viscum sp.)***: A well-known hemiparasitic plant that attaches to tree branches to extract water and nutrients while still performing photosynthesis.
- **Dodder *(Cuscuta sp.)***: Aholoparasitic plant that forms a dense tangle of thread-like stems on its host, completely relying on it for sustenance.
- ***Rafflesia arnoldii***: The largest flower in the world, which is a holoparasitic plant that lacks chlorophyll and depends entirely on its host for survival.

Fig 1: Different Parasitic Plants

Examples

- **Mistletoe *(Viscum sp.)***: Awell-known hemiparasitic plant that attaches to tree branches to extract water and nutrients while still performing photosynthesis.
- **Dodder *(Cuscuta sp.)***: Aholoparasitic plant that forms a dense tangle of thread-like stems on its host, completely relying on it for sustenance.
- ***Rafflesia arnoldii***: The largest flower in the world, which is a holoparasitic plant that lacks chlorophyll and depends entirely on its host for survival.

Our understanding of the evolution of plant parasitism has significantly increased due to the breakthroughs in our understanding of the connection between hosts and parasitic plants. Additionally, it is making it easier to create more potent control strategies in situations when these parasitic plants have turned into weeds in agriculture. The research paper by Masanga et al. (2021), by Bouwmeester (2021), in *Cuscuta campestris* and C. Kilimanjari (both either naturalized or native to East Africa), nicely illustrates the damage posed to agriculture by these weeds. These parasites have been shown to parasitize a wide variety of species from 13 different orders of Angiosperms. They have also been shown to pose a risk of spreading to major woody cash crops like mango, tea, and coffee.

Table 1: Biological features of some commonly found parasitic plants (Kokla & Melnyk, 2018). (HIFs: Haustoria Inducing Factors)

Species	Family	Parasitic Category	Haustoria Type	Infection Site	Photo synthetic Ability	HIFs	Vascular Connection
Phtheirospermum japonicum	Orobanchaceae	Facultative hemi parasite	Lateral	Root	Present	DMBQ, lignin polymers	Xylem
Triphysaria versicolor	Orobanchaceae	Facultative hemi parasite	Lateral	Root	Present	DMBQ, flavonoids	Xylem
Striga hermonthica	Orobanchaceae	Obligatory hemi parasite	Terminal	Root	Present	DMBQ, cell wall phenolics, lignin polymers, Strigolactones	Xylem
Cuscuta reflexa	Convolvulaceae	Obligatory hemi parasite	Lateral	Stem	Reduced	Volatile compounds	Xylem, Phloem
Phelipancheaegyptiaca	Orobanchaceae	Obligatory hemi parasite	Terminal	Root	Absent	Strigolactones	Xylem, Phloem

Ecology of Parasitic Plants

The ecology of parasitic plants centres on their interactions with host plants, their effects on plant communities, and their roles within ecosystems, which gives them a distinctive place in ecological dynamics.

- **Host-Parasite Interactions:** Such plants exhibit varying degrees of host specificity. Some are generalists, capable of parasitizing a wide range of host species, while others are specialists, targeting only specific hosts. Their impact on host plants may weaken them which can lead to reduced growth, lower reproductive success, and in severe cases, the death of the host.
- **Influence on Plant Communities:** Parasitic plants can both decrease and increase biodiversity. By weakening dominant species, they can reduce competition and create opportunities for other plants to establish, leading to greater diversity. In some ecosystems, parasitic plants act as keystone

species. Their presence and activity can shape the composition and diversity of plant communities, affecting the abundance and distribution of other species within the ecosystem.

- **Ecological Roles and Functions:** By influencing the health and productivity of host plants, parasitic plants play a role in nutrient cycling within ecosystems. For example, when a parasitized plant dies, it decomposes and returns nutrients to the soil, affecting soil fertility and plant succession. These plants can create microhabitats for various organisms. For instance, mistletoe, a hemiparasite, provides food and nesting sites for birds, which in turn disperse its seeds, creating a mutualistic relationship.

Parasitic plants affect natural and agricultural systems, and this can be considered a 'Dracula' effect (sucking the life out of an organism by reducing host productivity) or (robbing from the rich to feed the poor by extracting and redistributing resources among the plant community) (Twyford, 2018).

Life cycle of Parasitic Plants

Their life cycle involves a series of stages that allow them to locate, attach to, and extract nutrients from a host plant. This process can vary depending on whether the plant is a hemiparasite or a holoparasite. Parasitic plants must synchronise their lifecycle with their host to maximise fitness.The growth and development of parasitic plants is determined by a range of host attributes, such as root architecture and root spread, germination time and growth rate, and host physiochemistry. Host defence mechanisms also determine the success of a parasite (Twyford, 2018). The typical stages are explained below.

- **Seed Dispersal Mechanisms**: Seeds of parasitic plants are dispersed in various ways, including by wind, water, animals, or mechanical means. For example, mistletoe seeds are often dispersed by birds that eat the berries and later excrete the seeds onto branches.
- **Seed Dormancy and Germination**: Many parasitic plant seeds undergo a period of dormancy until they detect the presence of a suitable host. Their germination is often triggered by chemical signals released by potential host plants. For example, Striga seeds respond to strigolactones, which are chemicals exuded by the roots of host plants like maize or sorghum.
- **Host Location and Attachment**: After germination, the seedling grows towards the host plant, guided by chemical, light, or moisture gradients. Haustorium Formation takes place which penetrates the host's tissues and connects to its vascular system.Haustorium formation is an intriguing process, divided into three phases: (1) attachment to host tissue, (2) penetration of the host, and (3) connection to the host vasculature. (Brun et al., 2021).
- **Establishment and Growth**: Once the haustorium is established, the parasitic plant begins to draw resources from the host. In hemiparasites,

this is supplemented by photosynthesis, while holoparasites rely entirely on the host for survival and further it may spread to other nearby host plants.

- **Flowering, pollination & seed production**: The plant eventually flowers. Depending on the species, the flowers may be large and conspicuous or small and inconspicuous. Pollination can occur through various agents such as wind, insects, or birds, depending on the species.After successful pollination, seeds are produced and dispersed, restarting the cycle.
- **Impact on the Host**: Over time, the parasitic plant may weaken the host by depleting its resources, leading to reduced growth, lower reproductive success, and potentially the death of the host plant.

The life cycle of parasitic plants is closely tied to their ability to locate and exploit a suitable host. From seed dispersal to germination, host attachment, and reproduction, this parasitic lifestyle required the evolution of host detection, host attachment, host exploitation, and host defence suppression strategies (Bouwmeester et al., 2021), each stage is specialized to ensure the parasitic plant can successfully extract nutrients and propagate its species, often at the expense of the host plant's health.

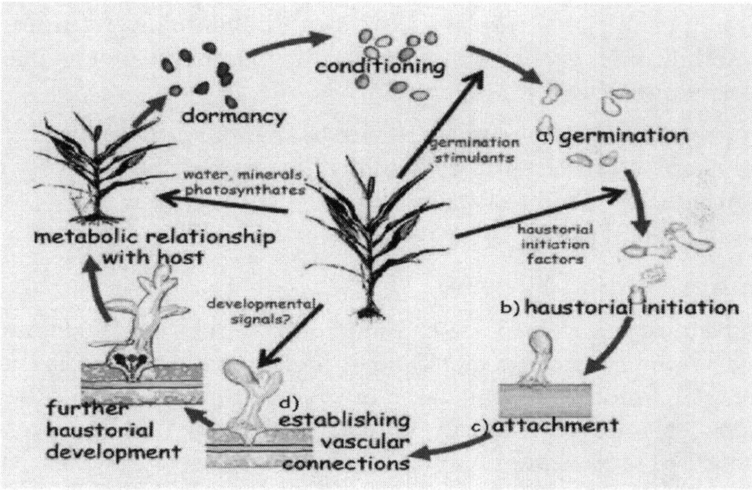

Fig 2: Life cycle of parasitic plants

Ecological and Environmental Impacts of Parasitic Plants

In addition, parasites control and occasionally have a significant impact on host population levels, trophic relationships, food webs, competition, biodiversity, and keystone species. These relationships imply that parasites play a crucial role in determining the organization of ecosystems and communities (Preston, 2010). Significant ecological and environmental effects of parasitic plants affect ecosystems' composition, efficiency, and biodiversity. Depending on the circumstances and the particular parasitic plant species involved, these effects might be either favorable or negative:

- Parasitic plants weaken the host, leading to reduced growth, lower reproductive success, and in severe cases, death. The decline of host plants can have cascading effects on the entire plant community.
- Parasitic plants can act as a selective pressure on host species, leading to the evolution of various defence mechanisms in hosts, such as changes in root structure, chemical defences, or altered growth patterns to avoid parasitism.
- By weakening or killing dominant plant species, parasitic plants can alter the composition of plant communities. This may reduce the dominance of certain species and allow less competitive species to establish and thrive, potentially increasing plant diversity.
- In some ecosystems, parasitic plants contribute to biodiversity by creating a more heterogeneous environment. Their presence can reduce the competitive advantage of dominant species, thus promoting a greater variety of plant species in the community.
- Parasitic plants can influence nutrient cycling within ecosystems. By extracting nutrients from their hosts and eventually returning them to the soil through decomposition, parasitic plants can alter soil fertility and nutrient availability for other plants.
- The presence of the plants can reduce the overall primary production of an ecosystem, especially if they significantly impact the health of dominant plant species. This reduction can affect the energy flow through the ecosystem and impact other trophic levels.

Impact on Agricultural Systems

One of the most troublesome pests of agricultural crops worldwide are parasitic plants, which number about 4000. To meet their nutritional needs, these plants take advantage of another plant's vascular system; roughly 90% of these species are hemiparasites, meaning they can still perform photosynthetic processes, while the remaining 390 species are holoparasites, meaning they must rely entirely on their host for all of their nutrients (Erdogan, 2021). Parasitic plants pose significant agricultural and economic challenges, particularly in regions where subsistence farming is common. Their ability to parasitize crop plants leads to reduced yields, increased production costs, and economic losses. Plants such as Striga (witch weed) and Orobanche (broomrape) are notorious agricultural pests that cause significant crop losses, particularly in developing regions. These plants can devastate fields of important crops like maize, sorghum, and legumes, leading to economic hardship for farmers and food insecurity in affected areas. Controlling parasitic plants in agriculture is difficult due to their often-cryptic early stages of development and their ability to produce large numbers of seeds. Effective management requires integrated approaches, including resistant crop varieties, chemical treatments, and cultural practices like crop rotation. Exploration of the agricultural and economic challenges associated with parasitic plants is:

Specific Crop Damage

Certain parasitic plants are known to target specific crops, for example, Striga (witchweed) affects cereals like maize, sorghum, and millet. It is especially problematic in sub-Saharan Africa, where it can cause yield losses of up to 100% in heavily infested fields. Orobanche (broomrape) targets legumes, sunflower, and tobacco, leading to significant crop losses, particularly in the Mediterranean, Middle East, and parts of Asia.

Economic Impact on Farmers

For smallholder farmers, the reduction in crop yields can be devastating, as it directly impacts their food security and income. In regions heavily affected by parasitic plants, entire communities can face economic hardship due to poor harvests and thus increased production costs.

Control Measures

Practices such as crop rotation, intercropping, and fallowing can help reduce the infestation of parasitic plants, but they require careful planning and may not be feasible for all farmers, especially those with limited land. In some cases, farmers resort to labour-intensive manual removal of parasitic plants, which increases labour costs and is not always effective in preventing re-infestation.Infestations by parasitic plants can also lead to increased post-harvest losses. For example, weakened plants may produce less robust seeds or fruits that are more susceptible to pests and diseases during storage.

Agricultural and Economic Challenges

Among the most destructive pests to agricultural crops globally are parasitic plants. Since the established parasite and host plant have a close physiological relationship that makes standard techniques of management ineffective, effective control measures are generally missing (Runyon, 2009). The intensive use of chemical herbicides to control parasitic plants can lead to soil degradation, reducing soil fertility and harming beneficial organisms in the soil. Over time, this can reduce the productivity of the land and Biodiversity Loss i.e. the spread of parasitic plants, particularly invasive species, can lead to the displacement of native plants and a reduction in plant diversity.Parasitic plants also reduce crop yields by removing essential nutrients from their host plants, causing slower growth, lower yields, and increased susceptibility to disease. They have similar nutritional requirements to their hosts and are less efficient at water use, leading to water stress in the host plants. They can also distort wood or alter growth patterns in trees, decreasing their value (Smith, 2024). This can have broader ecological impacts, including changes in habitat availability for wildlife and altered ecosystem functions.Parasitic plants, such as *Cuscuta reflexa* Roxb., have significant ecological and environmental impacts, both positive and negative. They affect Biodiversity and Ecosystem Balance by reducing the growth and vitality of their host plants, which can lead to decreased biodiversity, especially if they target dominant species. This can alter ecosystem structure by giving non-parasitized species a competitive advantage, leading to shifts in plant community composition which in turn directly affect Biodiversity and Ecosystem Balance. At the same time such plants are considered pests in agricultural systems as they reduce crop yields and affect food security. They also have a major Role in Succession and Habitat Change. They help in the transition of ecosystems by allowing less competitive species to thrive, promoting diversity in early stages of succession. Their participation in soil Nutrient Cycling is another matter of concern.Parasitic plants alter nutrient dynamics by affecting the nutrient flow in ecosystems. As they divert nutrients from their hosts, this can lead to reduced nutrient availability for other plants, affecting soil fertility and plant productivity.

A case study on *Cuscutareflexa*Roxb., The most widely growing Indian parasitic plant variety

This parasitic plant, which is a member of the Convolvulaceae family, is often referred to as dodder or amarbel. This plant is indigenous to Australia, Asia, and Africa's tropical and subtropical zones. It can be found in a number of African nations as well as Australia, Malaysia, Indonesia, Bangladesh, Nepal, Myanmar, Thailand, and Pakistan. It grows in both rain-fed and irrigated environments over most of India, including the states of Andhra Pradesh, Chhattisgarh, Gujarat, Orissa, West Bengal, and portions of Madhya Pradesh (Yadav, 2011).

Fig. 3: *Cuscuta reflexa* **Roxb. grown for study in University Campus**

Life cycle of *Cuscuta reflexa* Roxb. - It has a unique and interesting life cycle that revolves around parasitism. Here's an overview of its life cycle:

Stage 1, Seed Germination: The life cycle begins with the germination of seeds. The seeds of *Cuscuta reflexa* Roxb. are small and have a hard coat. When conditions are favourable, typically in the presence of moisture and warmth, the seed germinates. The seedling produces a thin, twining stem that searches for a host plant.

Stage 2, Host Location and Attachment: The seedling of *Cuscuta reflexa* Roxb. has a limited amount of stored energy and must find a host plant quickly to survive. Haustorium, begins to grow in a circular motion, attempting to make contact with a nearby host plant.Upon finding a suitable host, the stem wraps around it.

Stage 3, Parasitism: Once attached, *Cuscuta reflexa* Roxb. absorbs water, nutrients, and organic compounds from the host plant through the haustoria. This parasitic relationship allows it to survive and grow, even though it lacks chlorophyll and cannot perform photosynthesis on its own. The plant continues to grow and spread, producing more haustoria to invade additional parts of the host or even other nearby plants.

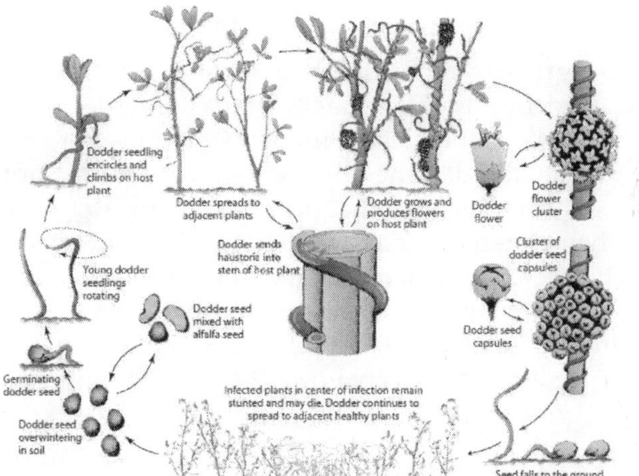

Fig 4: Dodder (*Cuscuta sp.*) disease cycle on a plant alfalfa

Stage 4, Growth and Flowering: As *Cuscuta reflexa* Roxb. grows, it forms a tangled mass of yellowish or orange stems, which can cover the host plant extensively. Eventually, the plant produces small, inconspicuous flowers that are typically white or pinkish in colour. These flowers are borne in clusters and are pollinated by insects.

Stage 5, Seed Production and Dispersal: After pollination, the flowers develop into small fruits containing seeds. The seeds are dispersed to the ground, where they can remain dormant until conditions are favourable for germination.The life cycle then repeats itself when the seeds germinate and the process begins anew.

Host plants suitable for *Cuscuta reflexa* Roxb.

It is a generalist parasite and can attach to a wide variety of host plants. As per study, it primarily parasitizes herbaceous plants, shrubs, and small trees. Some of the most common and suitable hosts noted during the study are:

- **Agricultural Crops:** Alfalfa (*Medicago sativa*), Tomato (*Solanum lycopersicum*), Chickpea (*Cicer arietinum*), Potato (*Solanum tuberosum*), Pea (*Pisum sativum*), Sugarcane (*Saccharum officinarum*).
- **Ornamental Plants:**Bougainvillea (*Bougainvillea spectabilis caryophyllales*), Chrysanthemum (*Dendranthema grandiflorum*), Dahlia (*Dahlia pinnata*), Marigold (*Tagetes spp.*).
- **Medicinal and Aromatic Plants:**Tulsi (*Ocimum sanctum*), Mint (*Mentha spp.*), Lemon balm (*Melissa officinalis*).
- **Wild Plants and Weeds:**Amaranthus (*Amaranthus cruentus*), Parthenium (*Parthenium hysterophorus*), Xanthium (*Xanthium strumarium*).
- **Shrubs and Small Trees:** Hibiscus (*Hibiscus rosasinensis*), Vinca (*Catharanthus roseus*), Bamboo (*Bambusa spp.*).

Its Ecological and Environmental Impacts

As a highly aggressive parasitic plant, *Cuscuta reflexa* Roxb. exerts profound ecological and environmental impacts that can destabilize ecosystems and agricultural landscapes. Ecologically, its parasitism leads to a reduction in plant diversity, as it targets a wide range of species, from wild plants to economically important crops. By drawing essential nutrients and water from its hosts, *Cuscuta reflexa* weakens them, often leading to reduced growth, lower reproductive success, or even death. This weakening or elimination of host plants can disrupt the balance of plant communities, potentially allowing invasive species to establish and spread more easily, further reducing biodiversity. Additionally, in natural ecosystems, the loss of key plant species can have cascading effects on the entire food web, impacting herbivores, pollinators, and other organisms that depend on those plants.

Environmentally, it poses a significant threat to agricultural productivity. Its ability to parasitize crops like alfalfa, tomato, and sugarcane leads to substantial yield losses, directly affecting food security and the livelihoods of farmers. The dense mats of the plant can smother crops, reducing photosynthesis and vigor, which not only impacts the current season's yield but can also weaken plants for future seasons. Furthermore, the management and control of *Cuscuta* infestations require considerable resources, including labor, herbicides, and other control measures, which can have additional environmental impacts, such as soil degradation and chemical runoff. The persistent and invasive nature of this plant also complicates efforts in land restoration and conservation, as it can invade and dominate disturbed areas, making it difficult to restore native vegetation and ecological balance.

Conclusion

In some cases, parasitic plants also play an important role in conservation and habitat restoration. By limiting the growth of invasive species, they help preserve native biodiversity and maintain ecosystem integrity. Parasitic plants drive co-evolutionary processes, influencing the evolution of defense mechanisms in host plants and fostering a complex ecological relationship between hosts and parasites. This can lead to the development of more resilient plant species.In conclusion, parasitic plants like *Cuscuta* have a dual role, both as disruptors and as contributors to ecosystem dynamics. Their management is crucial in agriculture but their ecological functions are also essential in maintaining biodiversity and ecosystem processes. Most naturally occurring parasitic plants should be considered an essential part of terrestrial ecosystems, contributing positively to biodiversity in a variety of organism groups. The available empirical data, which is still limited in terms of taxonomy and geographic scope, primarily highlights the effects of root hemiparasites and mistletoes on biodiversity (Tesitel, 2021).

References

Aukema, J. E. (2003). Vectors, viscin, and Viscaceae: mistletoes as parasites, mutualists, and resources. *Frontiers in Ecology and the Environment*, *1*(4), 212-219.

Bell, T. and Adams, M.A. (2011). Attack on all fronts: functional relationships between aerial and root parasitic plants and their woody hosts and consequences for ecosystems. Tree Physiol. 31:3–15.

Bouwmeester, H., Sinha, N., & Scholes, J. (2021). Parasitic plants: physiology, development, signalling, and ecosystem interactions. Plant Physiology, 185(4), 1267-1269.

Brun, G., Spallek, T., Simier, P., Delavault, P. (2021) Molecular actors of seed germination and haustoriogenesis in parasitic weeds. Plant Physiol 185: 1270–1281

Dobbertin, M., N., Hilker, M.,Rebetez, N.E., Zimmermann, T., Wohlgemuth, and. Rigling, A. (2005). The upward shift in altitude of pine mistletoe (Viscum album ssp. austriacum) in Switzerland – the result of climate warming? Int. J. Biometeorol. 50:40–47.

Erdogan P., (2021), Reviewed: 07 June 2021.Published: 13 July 2021DOI: 10.5772/intechopen.98760.

Heide J., Henning (2008). Parasitic flowering plants.doi:10.1163/ej.9789004167506.1-438 ISBN 9789047433590.

Kokla A., Melnyk C., Charles W., (2018). Developmental Biology. 442 (1): 53-59. doi: 10.1016/j.ydbio.2018.06.013 ISSN 0012-1606 PMID 29935146 S2CID 49394142.

Kokla, A., & Melnyk, C. W. (2018). Developing a thief: Haustoria formation in parasitic plants. Developmental biology, 442(1), 53-59.

Masanga J., Mwangi BN., Kibet W., (2021). Plant Physiol 185: 1457-1467 [PMC free article] [PubMed] [Google Scholar]

Nickrent, D.L. (2012). Parasitic plant genera and species. [online]. Available at http://www.parasiticplants.siu.edu/ParPlantNumbers.pdf [accessed 12 May 2015]

Parker, C., and Riches, C. R. (1993). Parasitic weeds of the world: biology and control (pp. xx+-332).

Pennings, S. C., & Callaway, R. M. (2002). Parasitic plants: parallels and contrasts with herbivores. Oecologia, 131(4), 479-489.

Pennings, S. C., and Callaway, R. M. (1996). Impact of a parasitic plant on the structure and dynamics of salt marsh vegetation. Ecology, 77(5), 1410-1419.

Press, M. C., & Phoenix, G. K. (2005). Impacts of parasitic plants on natural communities. New phytologist, 166(3), 737-751.

Preston, D. & Johnson, P. (2010) Ecological Consequences of Parasitism. Nature Education Knowledge 3(10):47Plant Physiol 185: 1374-1380 [PMC free article] [PubMed] [Google Scholar].

Runyon J.B., Tooker J.F., Mark C., (2009) DOI: 10.1007/978-1-4020-965498Source: In: E. Lichtfouse (ed.) Sustainable Agriculture Reviews 1. p. 123-136.

Smith D., Barkman T., (2024) volume 5, pages 112-120doi.org/10.1016/8978-0-12-822562-2.00332-7Academic Press, 2001, ISBN 9780123847201.

Tesitel J.; Knotkova K.; McLellan R. (2021) Plant Physiol. Apr; 185(4): 1309-1324doi: 10.1093/plphys/kiaa069PMCID: PMC8133642 | PMID: 33793868.

Twyford A. (2018) Mol. Biol. Evol. 2015; 32: 767-790Scopus (113) PubMed Crossref. Google Scholardoi.018.06.030

Westwood, J. (2024) Fact-checked by The Editors of Encyclopaedia Britannica Last Updated: Jul 31, Article History.

YadavR., Agarwala M., (2011).Phytochemical analysis of some medicinal plants. Journal of phytology, 3(12).Plants Research, (25):6017-6023 DOI:10.5897/JMPR11.869

CHAPTER 22

EDUCATION FOR SMART ENVIRONMENTAL MANAGEMENT AND PUBLIC AWARENESS

Dr. Gautami Bhatpahari

Assistant Professor of Psychology
Govt. D.B. Girls' PG College Raipur, Chhattisgarh

What is Environment?

'The surroundings or conditions in which a person, animal, or plant lives or operates' is environment.

The environment includes both living and nonliving things, such as soil, water, animals and plants, which adapt to their surroundings or what surrounds us is called the environment. It could be either living or non-living things. Physical, chemical and other natural forces are included in it. Living things exist in the environment they inhabit. They constantly interact with it and adapt to their environment.

In other words, environment *is* the *physical conditions* in which life exists. The environment is made up of components like animals, humans, sunlight, water, trees and air. *The planet has both* living and non-living components. *Plants, animals, and* trees *are living organisms. Thus, the environment* can be defined as *the* sum total of all living and non-living elements and their effects *on* human life. *The non*-living or *abiotic* elements *include water, land, sunlight, rocks*, and *air, while all* living or *biotic* elements *are animals,* plants, forests, and fisheries.

Types of Environments

The environment can be broken down into the following categories based on its basic structure:

- Physical/Abiotic Environment
- Biotic Environment
- Cultural Environment
- Social Environment

A physical or abiotic environment is divided into the following categories based on its physical characteristics and state:

1. Lithosphere, also known as solid earth
2. Hydrosphere, which contains water
3. Atmosphere, contains gaseous components

They can be further subdivided into smaller units based on various spatial scales, such as mountain, plateau, plain, lake, river, maritime, glacier, desert, and so on. The climatic conditions that provide specific habitats for biological communities, such as tropical, temperate, and polar environments, can also be viewed in terms of the physical environment

Biotic Environment: The biotic environment consists of living organisms, including both flora and fauna, with humans occupying a crucial position within it. Consequently, this ecosystem can be categorized into two main components: the floral environment and the faunal environment.

Floral Environment: This term typically denotes the variety of naturally occurring native plants. The name "Flora" originates from Latin, referring to the goddess of plants (floris means flower).

Faunal Environment: The faunal environment pertains to the habitat of animals. Fauna encompasses all the living creatures that are normally found in a specific area at a particular time

Thus, Flora encompasses all the vegetation, while fauna refers to all the wildlife within a specific area. The terms flora and fauna are used to describe any group of plants or animals. Each species interacts with and relies on other species, creating a foundation for an ecosystem. An ecosystem is a biological system made up of living and non-living things in an area and how they interact with each other. The word "eco" refers to a part of the world and "system" refers to coordination units. In an ecosystem, all components depend on each other. The cultural environment is an ecosystem in which society operates and is influenced by many factors such as socio-economic, technological and environmental factors. It can also refer to the customs and ideas of a group of people or society, which may be based on their culture, religion, place, country, language and other elements. Social Environment Social environments include the interactions, relationships and cultural systems that make up human societies. This type of environment plays an important role in the life of the individual and the whole, and influences the behavior, beliefs and social norms. Family environments influence socialization, emotional support and the transmission of cultural norms and values across generations. Educational environments include formal and informal environments where people acquire knowledge, skills and values. This includes schools and other educational institutions. Work environments include the places and conditions in which people engage in economic activities. From offices and factories to outdoor workplaces Community environments and social groups refer to the collective social context in which people interact with each other, interact and form social relationships.

What we get from environment?

Countless and endless things we get from the environment, for survival the basics are:

- **Air to breath**
- **Water to survive**
- **Food to grow**
- **Shelter to be safe**

Besides it

- Food and habitat for wildlife
- Improved waterway health.
- A more stable climate
- Oil and petroleum
- Minerals
- Vegetation
- Forests
- Rains
- Flood control
- Things good for our wellbeing-such as Concentrated minds &Reduced stress
- Immunity to resist and fight diseases.
- Sociability

Major threats to Environment

1. Dwindling Biodiversity

Human activities including logging, pollution, overfishing and urban development are driving a staggering loss of biodiversity. Global wildlife populations plummeted by 69% on average between 1970 and 2018, according to WWF's Living Planet Report 2022.

2. Fading Forests

Trees soak up planet-heating carbon dioxide and forests can lock away carbon for centuries, but they're disappearing at an alarming rate due to a combination of human activities.

3. Problem Plastic

It can take anywhere from 20 to 500 years for plastics to decompose and they can be found pretty much anywhere, from city streets to Antarctic sea ice and even drinking water. Not even the planet's tallest peak or the world's deepest ocean trench can escape plastic pollution.

4. Pollinators on the brink

Pollinators include bumblebees, wasps, moths and butterflies, as well as birds and small mammals, like bats. But many are facing threats from pesticides and crop monocultures, as well as the destruction of their habitats.

5. Bird populations under threat

Global bird populations face pressure from habitat destruction due to agriculture, development and logging, while seabirds are being impacted by overfishing.

All living things have been originated from the environmental growth and development. Environment is the essential part of life to exist and to continue the existence in this planet earth. To conserve and protect environment multiple and different facets of environmental education programs and awareness among public is needed.

Many governments Agencies and UNO and its different organizations have taken intense initiatives to conserve and protect environment some of them are listed below.

At the UNESCO's Tbilisi Conference (1977) the objectives of environmental education were

- **Awareness**: make people aware of the environment and the problems associated with it that may arise from improper use.
- **Education**: To help people gain experience and basic understanding of the environment and related problems.
- **Skills**: People need to acquire skills to understand and solve environmental problems.
- **Attitude**: Help people acquire environmental values and feelings, and encourage their participation in keeping the environment clean and safe.

Participation: opportunities at all levels to work to solve environmental issues

UNESCO ICEE at Tbilisi in (1977) laid down the guiding principles for environment protection are

The need for the hour is to consider the environment as a complete whole, that is, the natural, technological, social, economic, political, cultural, moral, historical and aesthetic environment.

The need is to have-

1. A continuous life process of environmental education—from pre-school to higher level and so as to non-formal education levels also.
2. An interdisciplinary approach to develop insight towards environmental problems.
3. Major issues of environment must be evaluated at from local, regional, national and international perspectives.
4. The complexities of environmental issues should be identified and recognized so as to foster critical thinking and problem-solving skills among people.
5. The cooperation of local, national and international is to be highlighted and be valued in addressing environmental problems.
6. Usage of available knowledge about the environment and the various available approaches of educating people about environment conservation.
7. People must be acquainted with the actual causes of environmental problems and the way these get expressed and must be encouraged to participate in prevention and resolving environmental problems.
8. Education at school level needs to link up aspects of sensitivity towards environment, knowledge of environment and problem-solving.
9. Stress on current as well as potential environmental situations.
10. All plans for social, economic and other kinds of growth and development must pay attention to environmental aspects.

Initiatives for public awareness and environmental education by Indian Government

The Government of India with Ministry of environment, forest and climate change (MOEFCC) has focused on environmental education and is undertaking programmes that involve public contribution to keep the environment safe. Few mentionable schemes are as follows:

1. Ministry's "Environmental Education, Awareness and Training (EEAT)" program

It aims raise people's awareness of the interdependence of humans and the environment at all levels and to provide them with the tools they need to conserve and enhance it. The main goals of this program, which was introduced in 1983–1984, were to encourage engagement from the public in environmental preservation and conservation and to raise environmental awareness among all societal segments. The following are the goals of the EEAT Scheme:

- To promote awareness towards environmental among all sections of the society.
- To foster environment education, especially in the non-formal system among different sections of the society.
- To facilitate development of education/training materials and aids in the formal education sector.
- To aid in environment education through existing educational/scientific/research institutions.
- To ensure training and manpower development for environment education, awareness and training.
- To encourage non-government organizations, mass media and other concerned organizations for promoting awareness about environmental issues among the people at all levels.
- To use media including films, audio, visual and print,, theatre, drama, advertisements, hoardings, posters, seminars, workshops, competitions, meetings etc. for spreading messages concerning environment and awareness.
- To mobilize people's participation for preservation and conservation of environment.

2. National Environment Awareness Campaign (NEAC)

The NEAC was launched in 1986 with the intention of raising national awareness of environmental issues. For the purpose of conducting awareness-raising and action-oriented activities, this campaign offers nominal financial assistance to educational institutions, research centers, colleges, women›s and youth organizations, NGOs, army units, government departments, and other organizations located throughout the nation. 34 Regional Resource Agencies (RRAs) have overseen and monitored NEAC activities. Over 11754 organizations from all over the nation took part in NEAC 2013–2014.

3. National Green Cops (NGC)

One of MOEFCC›s main initiatives to raise young children›s awareness of the environment is NGC. This program has demonstrated the value of grassroots efforts in raising public awareness of environmental issues. The program, which was introduced in 2001–2002, attempts to raise awareness among young people about environmental preservation and sustainable development. Within 14 years, the program created one of the biggest networks for conservation with over 1.0 million Eco clubs nationwide. This program gives schoolchildren opportunities to turn their ideas into creative action by giving them in-depth field experiences.

4. Swach Bharat Abhiyan

One of the major initiatives of the government towards a clean and safe environment is the Swach Bharat Abhiyan. The objectives of Swachh Bharat include the elimination of open defecation through the construction of family and community toilets and the creation of an accountable mechanism for monitoring the use of toilets. Led by the Government of India, the mission aims to achieve an Open Defeat-Free (ODF) India by October 2, 2019, the 150th anniversary of Mahatma Gandhi's birth, by building 12 million toilets in rural India at an estimated cost of Rs. 1.96 lakh crore ($30 billion). The mission includes two submissions: Swach Bharat Abhiyan ("Gramin" or rural), which is under the Ministry of Drinking Water and Sanitation; and Swach Bharat Abhiyan (urban), which comes under the Ministry of Housing and Urban Affairs. The mission includes ambassadors and activities like real-time national monitoring and updates from non-governmental organizations (NGOs) working towards their Swach Bharat ideas. Various public figures and brand ambassadors have been chosen by the government to spread the message. Thanks to massive volunteerism and government determination, overall national coverage of rural "household access to toilets" increased to 62.45 percent and usage to 91.29 percent in 2017.

Environment Protection

The Environment (Protection) Act was enacted in 1986 with the objective of providing for the protection and improvement of the environment. It empowers the Central Government to establish authorities [under section 3(3)] charged with the mandate of preventing environmental pollution in all its forms and to tackle specific environmental problems that are peculiar to different parts of the country. The Act was last amended in 1991..

Environmental Laws in India

1. The Water (Prevention and Control of Pollution) Act, 1974
2. The Water (Prevention and Control of Pollution) Rules, 1975
3. The Water (Prevention and Control of Pollution) Cess Act, 1977
4. The Water (Prevention and Control of Pollution) Cess Rules, 1978
5. The Air (Prevention and Control of Pollution) Act, 1981
6. The Air (Prevention and Control of Pollution) Rules, 1982
7. The Environment (Protection) Act, 1986
8. The Environment (Protection) Rules, 1986
9. Hazardous Wastes (Management and Handling) Rules, 1989
10. Manufacture, Storage and Import of Hazardous Chemical Rules, 1989
11. The Forest (Conservation) Act, 1980
12. The Forest (Conservation) Rules, 1981

13. The Wildlife Protection Act, 1972
14. The Wildlife (Transactions and Taxidermy) Rules, 1973
15. The Wildlife (Stock Declaration) Central Rules, 1973
16. The Wildlife (Protection) Licensing (Additional Matters for Consideration) Rules, 1983
17. The Wildlife (Protection) Rules, 1995
18. The Wildlife (Specified Plants - Conditions for Possession by Licensee) Rules, 1995
19. The Public Liability Insurance Act, 1991
20. The Public Liability Insurance Rules, 1991
21. The National Environment Tribunal Act, 1995
22. The National Environment Appellate Authority Act, 1997

Chhattisgarh government initiatives for creating awareness among people towards environment protection in accordance with its ministry and divisions-

1. **Chhattisgarh Rajya Van Vikas Nigam Limited**: In the year 1972, National Commission on Agriculture, in its report on "Production Forestry-Man Made Forests"expressed concern over fast depletion of forests and recommended- To enhance productivity of the forests by adopting Intensive Forest Management Technique and plantation of species of higher economic value.
2. **Chhattisgarh Environment Conservation Board**: The Chhattisgarh Environment Conservation Board (CECB) in Raipur was established in 2010. The CECB's functions are defined by the Air (Prevention and Control of Pollution) Act, 1981 and the Water (Prevention and Control of Pollution) Act, 1974.
3. **Chhattisgarh State Forest & Climate Change Department**: The State Centre for Climate Change and Knowledge Management Cell was established in August 2015 by State Government order. State Centre for Climate Change is the apex body to coordinate and implement the Chhattisgarh State Action Plan on Climate Change (CSAPCC).
4. **Chhattisgarh State Biodiversity Board**: Establishment date June 1, 2015, under the provisions of Section 63(1) of the 2002 act, Keeping in view the limited financial resources of the states, Forest Development Corporations should be formed to draw Institutional Finances

Notable Public Awareness Measures for Environment Conservation

1. **Kisan Vriksha Mitra Yojana** - objective is to increase the income of the beneficiaries by planting commercial species of trees on their lands and by ensuring buyback of the produce from them through the partner institutions and private companies. To increase the revenue of the government in the

form of additional tax by promoting cast and plywood based industries. To generate employment. To reduce the financial debt of the government by the participation of partner institutions in the tree plantation work.

2. **Plant for Mother**- Total number of plants planted under this campaign is 1966965
3. **Campa yojna** -The Compensatory Afforestation Fund Management and Planning Authority (CAMPA) is an organization established in 2004 by the Ministry of Environment and Forests, Government of India. The purpose of CAMPA is to manage money collected for compensatory afforestation, net present value (NPV), and other recoverable money. CAMPA also approves the use of forest land.
4. Life lifestyle for environment (A-Z sustainable lifestyle)

The concept of 'Lifestyle for the Environment (LiFE) was introduced by Hon'ble Prime Minister Shri Narendra Modi at COP 26 in Glasgow. He called upon the global community to rekindle a global pursuit of Lifestyle for Environment (LiFE). He urged to take LiFE forward as an international mass movement towards "mindful and deliberate utilization, instead of mindless and destructed consumption". The Mission emboldens the spirit of the P3 model, i.e., Pro Planet People, as it is premised on the basic principles of 'Lifestyle of the planet, for the planet and by the planet'

A. Alternatives to Plastics

B. Biodiversity Conservation

C. Circular Economy

D. Digital Consciousness

E. Energy Efficiency

F. Fuel Wise

G. Go Green

H. Healthy Planet Healthy Us

I. In Sensitive Rising Technology

J. Judicious Use of Resources

K. Knowledge Sharing

L. Landscaping

M. Micro Environment Management

N. Need Not Greed

O. Optimum Office

P. Prioritize Our Planet

Q. Quit Habits Antithetical to Environment

R. 6 Rs –Reduce, Reuse, Recycle, Recover, Re-Design and Re- Manufacture

S. Sustainable Agriculture

T. Transport

U. Uber urban

V. Value Your Culture

W. Waste Management

X. Xtravagance Avoidance

Y. Catch Them **Young**

Z. Zero Carbon Life Style

Indian Mind Set and Perception towards Environment Conservation

India is the land of rich cultural heritage where the nature place the important role, we celebrate everyday some or the other kind of or form of offerings, services and celebrations embedded with the natures; the plant Kingdom, The Water Kingdom and the animal kingdom are believed to be sacred and matter of worship. On the other hand we ruthlessly destroy the nature in the name of god, rituals and celebrations. This indicates the extreme perception and behavior towards the environment. Now the time has come to change this mind set and perception towards the environment or nature and take it as an objective to preserve and protect the environment. Besides the initiative and measures taken by international national and local governments, the accountability to conserve environment should be imposed on to each and every citizen irrespective of their position. The nature should be treated as a nature, a source of life and energy only, leaving behind the customs, rituals, offerings, celebration from nature. This will cultivate the new perception towards nature and individual of today will train themselves to be accountable and responsible for conserving and protecting the environment like a home.

References:

https://www.environment.nsw.gov.au/get-involved/sydney-nature/benefits-of-nature#:~:text=The%20natural%20environment%20gives%20us,floodwaters%2C%20and%20cool%20our%20streets

https://dspmuranchi.ac.in/pdf/Blog/satyapriya52dspmucomS12.pdf

http://3.https//www.google.com/search?q=Floral+environment+v.+Faunal+environment&rlz=1C1GCEA_enIN1042IN1042&oq=Floral+environment+v.+Faunal+environment&gs_ lcrp=EgZjaHJvbWUyBggAEEUYOTIHCAEQIRigATIHCAIQIRigATIHCAMQIRigATIHCAQQIRigAdIBCDI1NTRqMGo3qAIIsAIB&sourceid=chrome&ie=UTF-8

https://www.google.com/search?q=ecosystem&rlz=1C1GCEA_enIN1042IN1042&oq=ecosystem&gs_lcrp=EgZjaHJvbWUyEQgAEE UY-

ORhDGLEDGIAEGIoFMgwIARAAGEMYgAQYigUyDAgCEAAYFBiH AhiABDIPCAMQABhDGLEDGIAEGIoFMg0IBBAuGK8BGMcBGIAEMg8I-BRA AGEMYsQMYgAQYigUyBwgGEAAYgAQyBggHEEUYPNIBCDc0 NTZqMGo3qAIIsAIB&sourceid=chrome&ie=UTF-8

https://sigmaearth.com/types-of-environment-a-complete-guide/

https://forest.cg.gov.in/posts/flagship-schemes

https://forest.cg.gov.in/maakenaam/index.html

https://drive.google.com/file/d/1zNlSNn8lPY6HygxmHuixrWr7tScuSViv/view

https://forest.cg.gov.in/

https://moef.gov.in/uploads/2017/06/eprotect_act_1986.pdf

https://missionlife-moefcc.nic.in/assets/flipbook/A-Z-Sustainable-Lifestyle-ebook/A-Z-Sustainable-Lifestyle-ebook.html#page/20

https://iica.nic.in/images/presentation/Environmental%20Laws.pdf

https://rvvn.cgstate.gov.in/hi

CHAPTER 23

SUSTAINABLE RESOURCES MANAGEMENT

Dr. Sangeeta Jha Aditya[1] and Dr. Razia Sultana[2]

[1]*Department of Economics, Govt. Arts and Commerce Girls' College, Raipur C.G.*
sangaditya222@gmail.com

[2]*Department of Zoology, Shri Kuleshwar Mahadev Govt. College Gobra-Nawapara Raipur, CG.*
raziasultana83ster@gmail.com

Abstract

Sustainable resource management is essential for balancing environmental conservation, economic development, and social equity. This chapter provides a comprehensive overview of sustainable resource management principles, strategies, and challenges. It explores key concepts such as the sustainable use of resources, integrated management approaches, and the role of innovation and technology in achieving sustainability goals. The chapter examines case studies from various sectors, including water, forests, and agriculture, highlighting successful practices and lessons learned. It also addresses the complexities and obstacles faced in implementing sustainable management, including policy, economic, and social barriers. The chapter concludes with recommendations for enhancing sustainability practices and ensuring effective resource management in the face of growing global demands and environmental pressures.

Keywords: Sustainable resource management, environmental conservation, economic development, integrated management, innovation, case studies, water management, forest management, agriculture, policy challenges.

Introduction

Sustainable resource management is a critical component of ensuring long-term ecological balance and socio-economic stability. This chapter explores the principles, practices, and challenges of managing natural resources in a way that meets present needs without compromising the ability of future generations to meet theirs. Sustainable resource management is vital for ensuring that natural resources continue to provide essential services to humanity. By adopting integrated, adaptive, and equitable management strategies, we can address the complex challenges posed by environmental degradation, climate change, and socio-economic inequalities. The concept of sustainable resource management has gained prominence in recent decades, particularly as the global population has grown and the demand for natural resources has intensified. Sustainable resource management seeks to harmonize the exploitation of natural resources with the ecological, economic, and social needs of the present and future generations. One of the most significant challenges in sustainable resource management is climate change. As global temperatures rise, ecosystems are disrupted, leading to shifts in resource availability. For example, freshwater supplies are increasingly threatened by melting glaciers and altered precipitation patterns, making sustainable water management a critical area of focus.

Historical Context and Evolution of Sustainability

The idea of sustainability has ancient roots, with early civilizations recognizing the importance of living in harmony with nature. However, the modern concept of sustainability began to take shape in the late 20th century, particularly with the environmental movement of the 1960s and 1970s, which highlighted the detrimental impacts of industrialization on the planet. The concept was formally articulated in 1987 by the Brundtland Commission in the report titled "Our Common Future." The report defined sustainable development as "development that meets the needs of the present without compromising the ability of future generations to meet their own needs." This definition emphasized two key ideas, **Intergenerational Equity** to ensure that future generations inherit a world capable of supporting their needs and **holistic Approach to r**ecognize environmental, economic, and social factors are interconnected and must be considered together in decision-making.

Principles of Sustainable Resource Management

The principle of intergenerational equity is foundational to sustainable resource management. It requires that current resource use does not compromise the ability of future generations to meet their needs. This principle is closely linked to the ethical considerations of resource allocation, where decisions must balance the needs of today's population with the rights of future generations. Integrated Resource Management (IRM) represents a holistic approach that considers the interconnectedness of various resources. For example, managing a watershed involves not just water but also land use, forestry, and agriculture. By considering these elements together, IRM promotes a balanced approach to resource use that can mitigate negative impacts on the environment. Following are the key principles guide sustainable resource management, each of which is essential to achieving sustainability.

Intergenerational Equity

Intergenerational equity is the principle that today's resource use should not diminish the capacity of future generations to meet their needs. This concept emphasizes the ethical responsibility of current generations to manage resources in a way that does not compromise the opportunities and well-being of those who will come after us. In includes resource conservation, Long-term planning and ethical responsibility. **Resource Conservation** advocates for the sustainable use of both renewable and non-renewable resources. For renewable resources, such as forests and fisheries, this means managing them at a rate that allows for natural regeneration. For non-renewable resources, such as fossil fuels and minerals, it suggests minimizing use, improving efficiency, and investing in alternative technologies. **Long-Term Planning is i**ntergenerational equity requires a shift from short-term, profit-driven exploitation of resources to long-term planning and management strategies that consider the needs of future populations. And e**thical responsibility** is rooted in the belief that current generations have an ethical obligation to preserve the planet's resources for future generations. This includes protecting the environment, maintaining biodiversity, and preventing the depletion of critical natural resources.

Conservation of Biodiversity

It is the principle that recognizes the intrinsic value of the variety of life on Earth and its crucial role in maintaining the resilience and functionality of ecosystems. Biodiversity is essential for ecosystem stability, providing services such as pollination, water purification, climate regulation, and soil fertility. It includes e**cosystem Integrity.** Healthy ecosystems with high biodiversity are more resilient to environmental changes and disruptions. By conserving biodiversity, we maintain the integrity and functionality of ecosystems, which in turn supports sustainable resource management. Second is s**pecies protection** focus on protecting species from extinction, preserving genetic diversity, and maintaining habitats. This includes the creation of protected areas, wildlife corridors, and the restoration of degraded ecosystems. Finally, s**ustainable use of ecosystems** is the principle of sustainable use involves managing ecosystems in ways that do not lead to long-term depletion of biodiversity. This includes practices like sustainable forestry, agriculture, and fishing, which aim to balance resource extraction with the preservation of ecosystem health.

Efficiency and Equity

The principles of **efficiency and equity** address the need for fair distribution of resources and the efficient use of those resources to minimize waste and environmental impact. It includes efficient use of resources, equitable access and distribution and addressing inequalities. Efficiency involves maximizing the utility and benefits derived from resources while minimizing waste and environmental damage. This includes adopting technologies and practices that reduce resource consumption, increase productivity, and promote the circular economy, where materials are reused, recycled, and repurposed. Equity refers to the fair and just distribution of resources across different segments of society. It involves ensuring that all individuals and communities have access to the resources they need to meet their basic needs and improve their quality of life. This principle also addresses issues of social justice, recognizing that marginalized communities often bear the brunt of environmental degradation while having limited access to resources. Further, sustainable resource management must account for existing inequalities in resource access and consumption. Policies and practices should aim to reduce disparities, ensuring that vulnerable and disadvantaged populations are not disproportionately affected by resource scarcity or environmental degradation.

Adaptive Management

Adaptive management is a dynamic approach to resource management that recognizes the complexity and uncertainty inherent in natural systems. It involves continually monitoring and adjusting management strategies in response to new information and changing conditions. It includes flexibility and learning, incorporating scientific research, integrating traditional knowledge, case study and learning by doing. Adaptive management emphasizes the need for flexibility in management practices. It is based on the idea that our understanding of ecosystems and the impacts of human activities is incomplete and constantly evolving. By adopting a flexible approach, resource managers can respond to new challenges and opportunities as they arise. Adaptive management relies on ongoing scientific research and monitoring to inform decision-making. This includes studying ecosystem processes, the impacts of management actions, and the effects of external factors such as climate change. In many cases, traditional ecological knowledge, held by indigenous and local communities, provides valuable insights into sustainable resource management. Adaptive management seeks to integrate this knowledge with scientific research to create more effective and culturally appropriate management strategies. Adaptive management often involves implementing pilot projects or experimental approaches on a small scale, learning from the outcomes, and then scaling up successful strategies. For example, adaptive management has been used in fisheries management, where real-time data on fish populations are used to adjust quotas and harvesting practices.

Polluter Pays Principle

The **polluter pays principle** is an environmental policy principle that requires those who cause environmental harm to bear the costs of managing and mitigating that harm. This principle incentivizes businesses, industries, and individuals to reduce pollution and adopt more sustainable practices. It involves internalizing environmental costs, economic instruments, and incentives for sustainable practices. Traditionally, many environmental costs, such as pollution and habitat destruction, have been externalized, meaning that society as a whole bears the burden. The polluter pays principle seeks to internalize these costs, ensuring that those who benefit from resource extraction or industrial activities also pay for the environmental impacts. The principle is often implemented through economic instruments such as taxes, fines, and fees. For example, a carbon tax imposes a financial cost on the emission of greenhouse gases, encouraging businesses to reduce their carbon footprint. Further, by holding polluters accountable for environmental damage, the polluter pays principle creates incentives for adopting cleaner technologies, improving efficiency, and reducing waste. This can lead to innovation in sustainable resource management practices.

Precautionary Principle

The **precautionary principle** is a risk management approach that advocates for proactive action in the face of uncertainty. It suggests that when there is a risk of significant or irreversible harm to the environment or human health, the lack of full scientific certainty should not be used as a reason to delay action. It includes prevention of environmental harm, policy and regulation, balance innovation and safety. The precautionary principle is particularly relevant in situations where the potential impacts of an activity are not fully understood, such as the introduction of new technologies, chemicals, or industrial practices. It promotes taking preventive measures to avoid harm, even if the evidence of risk is not conclusive. The precautionary principle is often applied in environmental policy and regulation, guiding decisions on issues such as biodiversity conservation, chemical safety, and climate change. For example, it has been used to justify regulations on the use of pesticides, genetically modified organisms (GMOs), and the management of hazardous waste. While the precautionary principle encourages caution, it also requires a balance between innovation and safety. It does not call for halting progress but rather for carefully considering the potential risks and benefits of new developments and making informed decisions that prioritize long-term sustainability.

Concept of Environmental, economical and social sustainability

Sustainability is a multifaceted concept that has gained significant attention in recent decades, particularly in the context of environmental conservation, economic development, and social equity. At its core, sustainability is about meeting the needs of the present without compromising the ability of future generations to meet their own needs. This concept, which originated in environmental discourse, has evolved to encompass economic, social, and cultural dimensions, making it a guiding principle for global development.

Sustainability is often conceptualized as resting on three interdependent pillars: environmental, economic, and social sustainability. These pillars provide a framework for understanding and addressing the complex challenges facing humanity. Environmental sustainability involves the responsible management of natural resources to ensure that ecosystems remain healthy and productive over time. It focuses on **Conservation of Natural Resources, biodiversity preservation and pollution control. Conservation includes** use of renewable resources (like forests, fisheries, and freshwater) at a rate that allows them to regenerate. For non-renewable resources (like minerals and fossil fuels), it involves finding alternatives and reducing reliance. Secondly, b**iodiversity preservation is crucial for maintaining** the diversity, stability, and resilience of life on Earth, and ecosystem. Finally, uncontrolled human development and population generates undesirable pollutants to the environment that effect health and environment. Minimization of waste and emissions can reduce the impact on air, water, and soil quality. Examples of environmental sustainability practices include sustainable agriculture, renewable energy production, and ecosystem restoration.

Economic sustainability refers to practices that support long-term economic growth without negatively impacting social, environmental, and cultural aspects of the community. It involves e**fficient use of resource, m**aximizing the productivity of resources while minimizing waste and environmental impact; i**nnovation and Technology** for developing and adopting new technologies that reduce resource consumption and environmental impact; and s**ustainable Economic Models** for Promoting economic activities that are financially viable in the long term and that do not deplete natural resources or harm the environment. Economic sustainability is closely linked to concepts like the circular economy, where resources are reused, recycled, and repurposed to minimize waste and extend the life cycle of products.

Social sustainability is about creating an equitable society that meets the needs of all its members, particularly the most vulnerable, and ensures that future generations can thrive. It focuses on equity and justice. **Equity and Justice** ensures fair distribution of resources and opportunities, addressing inequalities, and promoting social inclusion. **Community Development** can be strengthening by providing access to education, healthcare, and employment, and by fostering social cohesion. In addition, Protecting and promoting cultural diversity and heritage, which are integral to community identity and resilience. Social sustainability also includes fostering institutions and governance structures that are transparent, accountable, and responsive to the needs of all stakeholders.

These three pillars of sustainability—environmental, economic, and social—are deeply interconnected. Actions in one area can significantly impact the others, either positively or negatively. For example deforestation or pollution can lead to the loss of livelihoods for communities dependent on natural resources, exacerbating poverty and social tensions. In addition, industrial growth that prioritizes short-term economic gains without considering environmental costs can lead to resource depletion, loss of biodiversity, and long-term economic instability. And communities with strong social support systems are often better able to implement and maintain sustainable environmental practices. A sustainable approach to development recognizes these linkages and seeks to balance the needs of the environment, economy, and society in a way that promotes long-term well-being for all.

Strategies for Sustainable Resource Management

Sustainable resource management involves a range of strategies designed to ensure that natural resources are used in a way that maintains their availability and quality for future generations while supporting current economic and social needs. Below are detailed explanations of several key strategies that are widely adopted to achieve sustainable resource management.

Integrated Resource Management (IRM): Integrated Resource Management (IRM) is a holistic approach that considers the interconnections between different natural resources and the various sectors that depend on them. Rather than managing resources in isolation, IRM recognizes the complex interactions within ecosystems and across different resource uses. IRM involves the management of water, land, forests, and other resources across different sectors such as, agriculture, industry, and urban development. For example, in watershed management, IRM considers how land use (such as agriculture or urbanization) affects water quality and availability, as well as how water management impacts other sectors. IRM seeks to balance the often-competing interests of different stakeholders, including local communities, businesses, and government agencies. This approach requires collaboration and negotiation to develop management plans that meet the needs of all parties while ensuring the sustainability of the resources. For example, in forest management, IRM might involve balancing timber production with conservation efforts, water regulation, and recreational use. By considering the full range of ecosystem services provided by forests, IRM helps to ensure that these resources are managed sustainably.

Ecosystem-Based Management (EBM): that focuses on managing resources within the context of entire ecosystems rather than targeting specific resources or species. EBM recognizes that ecosystems are complex, dynamic systems where different components—such as plants, animals, water, and soil—interact in ways that are often not fully understood. The EBM, aims to maintain the health, function, and resilience of ecosystems. This involves considering the cumulative impacts of human activities on ecosystems and managing them in a way that preserves their ability to provide essential services, such as clean water, air, and food. The EBM also encourages decision-makers to consider the broader ecological, social, and economic implications of resource management actions. For example, in marine environments, EBM might involve managing fisheries, coastal development, and pollution together to protect the overall health of the marine ecosystem. For example, the management of coral reefs using EBM might involve regulating fishing practices, controlling land-based pollution, and protecting critical habitats to ensure the long-term sustainability of the reef ecosystem and the services it provides, such as tourism and fisheries.

Community-Based Resource Management (CBRM): involves the active participation of local communities in managing the natural resources they depend on. This approach recognizes the valuable knowledge and vested interests that local people have in the sustainable use of their environment. CBRM empowers communities to take ownership of the resources in their area. It involves local people in decision-making processes, ensuring that management plans reflect their needs, values, and traditional knowledge. CBRM often focuses on creating sustainable livelihoods for communities, reducing their reliance on unsustainable practices such as overfishing or deforestation. This might involve promoting alternative income sources, such as eco-tourism or sustainable agriculture that are less harmful to the environment. In many parts of the world, CBRM is used in the management of fisheries. Local fishers are involved in setting catch limits, protecting breeding grounds, and enforcing regulations, leading to more sustainable fishing practices and improved fish stocks over time.

Technological Innovations: Technological innovations play a critical role in advancing sustainable resource management by providing new tools and techniques for monitoring, managing, and conserving natural resources. Precision agriculture uses technologies such as GPS, remote sensing, and data analytics to optimize farming practices. By applying inputs like water, fertilizers, and pesticides more precisely, farmers can reduce waste, lower costs, and minimize environmental impacts while maintaining or increasing crop yields. The development and deployment of renewable energy technologies, such as solar, wind, and hydropower, are essential for reducing dependence on fossil fuels, which are finite and contribute to environmental degradation. The adoption of renewable energy supports the sustainable use of energy resources and helps mitigate climate change. Further, innovations in water management, such as drip irrigation, rainwater harvesting, and wastewater recycling, are crucial for conserving water resources, particularly in arid and semi-arid regions. These technologies improve water-use efficiency, reduce waste, and ensure that water remains available for both human consumption and ecosystem needs. Geographic Information Systems (GIS) and remote sensing technologies are widely used in forest management to monitor deforestation, map forest cover, and assess the health of ecosystems. These tools provide valuable data that support informed decision-making and the sustainable management of forest resources.

Market-Based Approaches: Market-based approaches to sustainable resource management involve the use of economic incentives and disincentives to encourage the sustainable use of natural resources. These approaches leverage market mechanisms to align economic activities with environmental goals. It includes Payment for Ecosystem Services (PES). PES schemes involve compensating landowners or resource users for managing their land in ways that provide environmental benefits, such as carbon sequestration, water purification, or biodiversity conservation. For example, farmers might be paid to maintain forest cover on their land to protect a watershed. Secondly, Tradable permit systems, such as cap-and-trade schemes for carbon emissions, set a limit on the total amount of a resource that can be used or pollution that can be emitted. Companies or individuals can buy and sell permits within this cap, creating financial incentives to reduce resource use or emissions. Eco-labelling and certification programs, such as Fair Trade or organic certification, provide consumers with information about the environmental and social impacts of products. By choosing certified products, consumers can support sustainable practices and encourage companies to adopt more environmentally friendly methods. The carbon offset market allows businesses and individuals to purchase credits representing the reduction or removal of carbon emissions elsewhere. This market-based approach incentivizes investment in renewable energy, reforestation, and other carbon-reducing projects, contributing to global efforts to combat climate change.

Regulatory and Policy Frameworks: Regulatory and policy frameworks are essential for setting the rules and standards that guide sustainable resource management. Governments play a crucial role in developing and enforcing policies that promote the sustainable use of natural resources. Environmental regulations set limits on resource use and pollution, protecting ecosystems and human health. These regulations may include restrictions on land use, water withdrawals, emissions, and waste disposal. Compliance with these regulations is enforced through monitoring, inspections, and penalties for violations. Governments can implement policies that promote sustainable development, such as subsidies for renewable energy, tax incentives for conservation practices, and support for research and innovation in sustainable technologies. These policies help to create an enabling environment for sustainable resource management. Many environmental challenges, such as climate change, deforestation, and overfishing, are global in nature and require international cooperation. It also involves the Treaties and agreements, such as the Paris Agreement on climate change and the Convention on Biological Diversity, set international standards and goals for sustainable resource management. The European Union's Common Agricultural Policy (CAP) includes provisions for sustainable agriculture, such as payments to farmers who adopt environmentally friendly practices. This policy framework encourages sustainable land management and the protection of natural resources across member states.

Restoration and Rehabilitation: Restoration and rehabilitation strategies involve repairing and restoring ecosystems that have been degraded or damaged by human activities. These strategies are essential for recovering the health and functionality of ecosystems, ensuring that they continue to provide valuable services. Restoration efforts aim to return ecosystems to their original state or to a state where they can function sustainably. This may involve reforestation, wetland restoration, or the reintroduction of native species. The Restoration projects often focus on areas that have been heavily impacted by activities such as mining, agriculture, or urban development. Further, rehabilitation efforts focus on improving the productivity and ecological integrity of degraded land. This may involve practices such as soil conservation, erosion control, and the planting of cover crops. The Land rehabilitation is particularly important in areas affected by desertification, deforestation, and poor agricultural practices. The restoration of the Florida Everglades is one of the largest and most ambitious ecosystem restoration projects in the world. The project involves restoring natural water flows, removing invasive species, and rehabilitating wetlands to protect this unique and valuable ecosystem.

Challenges in Sustainable Resource Management:

Sustainable resource management faces numerous challenges that arise from the complexity of balancing environmental, economic, and social goals. These challenges can hinder efforts to ensure that natural resources are used responsibly and remain available for future generations. Below are some of the key challenges in sustainable resource management:

1. **Resource Scarcity and Overexploitation:** It includes resource scarcity, increased demand, depletion of non-renewable resources, and overfishing and deforestation.
2. **Climate change** is one of the most significant challenges to sustainable resource management. It exacerbates resource scarcity, disrupts ecosystems, and increases the frequency and severity of natural disasters, making it more difficult to manage resources sustainably.
3. **Social and economic inequities** can hinder sustainable resource management by creating disparities in resource access, use, and benefits. Inequities can lead to conflict, overexploitation, and unsustainable practices, particularly in regions where marginalized communities have limited rights or access to resources.
4. **Governance and Policy Challenges Weak governance and inadequate policies** can undermine sustainable resource management efforts. Effective governance is essential for enforcing regulations, protecting resources, and ensuring that management practices are sustainable and equitable.
5. **Technological and Knowledge Gaps** pose significant challenges to sustainable resource management. Without the necessary technologies and knowledge, it is difficult to monitor resource use, assess environmental impacts, and implement sustainable practices.
6. **Population Growth and Urbanization** increase the demand for resources and contribute to environmental degradation. As more people move to cities, the pressure on surrounding natural resources intensifies, leading to challenges in managing resources sustainably.
7. **Globalization and Market Pressures** can drive unsustainable resource use by encouraging the overexploitation of resources for short-term economic gain. The global demand for natural resources, coupled with the pursuit of economic growth, often leads to environmental degradation and social inequities.

The Role of Innovation and Governance and future Sustainability

Innovation and governance play crucial roles in advancing sustainability. Advances in technology, such as renewable energy, sustainable agriculture practices, and resource-efficient manufacturing, are essential for reducing environmental impact while supporting economic growth. Effective policies, regulations, and governance structures are necessary to guide sustainable practices, enforce environmental protections, and promote equitable development. International agreements like the Paris Agreement on climate change, the United Nations Sustainable Development Goals (SDGs), and national policies on renewable energy and conservation are examples of efforts to institutionalize sustainability.

As the global community continues to face the challenges of the 21st century, the concept of sustainability will remain central to efforts to create a more resilient and equitable world. The future of sustainability will likely involve **global cooperation, a**ddressing global challenges like climate change and biodiversity loss will require unprecedented levels of international cooperation. In addition, Communities around the world will need to adopt and adapt sustainable practices that are tailored to their unique environmental and socio-economic contexts. Future approaches to sustainability will increasingly recognize the interconnectedness of environmental, economic, and social systems, promoting solutions that address all three dimensions simultaneously.

Brief Case Study: Sustainable Water Management in Arid Regions – The Case of Israel

Israel, a country with limited natural freshwater resources and an arid climate, has become a global leader in sustainable water management. The country faces significant water scarcity due to low rainfall, high evaporation rates, and increasing demand from agriculture, industry, and urban populations. Despite these challenges, Israel has developed innovative strategies and technologies to manage its water resources sustainably such as desalination, water recycling and reuse, efficient irrigation techniques, public awareness and policy measures. The country invested heavily on recycling and desalination of water. Israel's integrated water management approach has enabled the country to overcome severe water scarcity and ensure a stable water supply for its population and economy. The success of Israel's water management strategies has made the country a model for other arid and semi-arid regions facing similar challenges. This case study provides valuable insights for other countries facing water scarcity challenges.

Brief Case Study: Forest Management in the Amazon Rainforest

The Amazon rainforest, often referred to as the "lungs of the Earth," is the largest tropical rainforest in the world, spanning several countries in South America, with the majority located in Brazil. It is home to unparalleled biodiversity and plays a crucial role in regulating the global climate. However, the Amazon faces significant threats from deforestation, illegal logging, agricultural expansion, and climate change. Sustainable forest management in the Amazon has become essential to preserving this vital ecosystem while balancing economic needs. The conservative strategy includes **Community-Based Forest Management (CBFM), 50% Protected Areas and Conservation Units,** Sustainable logging practices have been introduced in the Amazon to reduce the environmental impact of timber extraction, d**eforestation Monitoring and Enforcement. The impacts involve reduction in deforestation rates, empowerment of indigenous and local communities, promotion of Biodiversity and Ecosystem Services.** Sustainable forest management in the Amazon Rainforest is a complex but essential endeavor. Strategies such as community-based management, protected areas, sustainable logging, and deforestation monitoring have shown positive results in reducing deforestation and empowering local communities. However, continued vigilance, stronger enforcement, and international cooperation are necessary to overcome ongoing challenges and ensure the long-term preservation of the Amazon and its critical global environmental role.

Brief Case Study: Conservation Agriculture in Zambia

Zambia, a country in Sub-Saharan Africa, relies heavily on agriculture, with the majority of its population engaged in smallholder farming. However, the sector faces significant challenges such as soil degradation, low productivity, and the impacts of climate change, including erratic rainfall. To address these issues, conservation agriculture (CA) has been promoted as a sustainable farming practice in Zambia since the early 2000s. **Strategies Implemented was minimum soil disturbance, permanent soil cover, crop Rotation and intercropping, Farmer Training and Support. The strategy results in increased Crop Yields, improved soil health,** Conservation agriculture has led to better soil health, with enhanced soil structure, increased organic matter, and improved water retention. This has reduced soil erosion and degradation, ensuring long-term agricultural productivity and e**nhanced climate resilience.** Despite its benefits, the widespread adoption of conservation agriculture in Zambia has been hindered by barriers such as initial labor demands, limited access to inputs, and resistance to changing traditional farming methods. Continuous support in terms of training, input supply, and market access is needed to overcome these challenges. Conservation agriculture in Zambia offers a sustainable solution to the challenges of low productivity, soil degradation, and climate variability. By improving soil health and increasing crop yields, CA has contributed to food security and resilience among smallholder farmers. However, ongoing efforts are required to scale up these practices and ensure their long-term sustainability across the country.

Conclusion

Sustainable resource management requires a diverse set of strategies that address the complexity of managing natural resources in a way that balances environmental, economic, and social needs. Integrated Resource Management, Ecosystem-Based Management, Community-Based Resource Management, technological innovations, market-based approaches, regulatory and policy frameworks, and restoration and rehabilitation efforts all play crucial roles in ensuring the sustainable use of resources. By applying these strategies, societies can work towards a future where natural resources are managed responsibly, supporting both human well-being and the health of the planet. By embracing intergenerational equity, biodiversity conservation, efficiency, equity, adaptive management, the polluter pays principle, and the precautionary principle, societies can address the complex challenges of resource depletion, environmental degradation, and social inequities. By embracing the principles of sustainability, humanity can work towards a future that is both prosperous and harmonious, ensuring that the planet's resources are preserved and shared fairly among all its inhabitants. This detailed explanation of sustainability provides a comprehensive overview of the concept, its evolution, and its application in various contexts. It also highlights the challenges and opportunities associated with achieving sustainability on a global scale.

References

Barbier, E. B., & Burgess, J. C. (2021). The economics of sustainable resource management. In Handbook of Environmental Economics (Vol. 3, pp. 345-374). Elsevier.

Kates, R. W., Parris, T. M., & Leiserowitz, A. A. (2005). What is sustainable development? Goals, indicators, values, and practice. Environment: Science and Policy for Sustainable Development, 47(3), 8-21.

MEA. (2005). Millennium Ecosystem Assessment: Synthesis Report. Island Press.

Ostrom, E. (2009). A general framework for analyzing sustainability of social-ecological systems. Science, 325(5939), 419-422.

Rockström, J., & Sukhdev, P. (2016). Sustainable resource management for future food security. In Global Food Security: Challenges and Strategies (pp. 54-75). Springer.

UNEP. (2011). Towards a Green Economy: Pathways to Sustainable Development and Poverty Eradication. United Nations Environment Programme.

Wilson, E. O. (2016). Half-Earth: Our Planet's Fight for Life. Liveright Publishing Corporation.

अध्याय 24

सतत संसाधन प्रबंधन

नीलिमा पाण्डेय, सहायक प्राध्यापक
neelima.aashish@gmail.com

सारांश

सतत संसाधन प्रबंधन (सस्टेनेबल रिसोर्स मैनेजमेंट) का उद्देश्य प्राकृतिक संसाधनों का इस प्रकार प्रबंधन करना है कि वे वर्तमान पीढ़ी की आवश्यकताओं को पूरा करते हुए भविष्य की पीढ़ियों के लिए भी उपलब्ध रहें। यह प्रबंधन प्राकृतिक संसाधनों जैसे जल, भूमि, वनों, खनिजों और ऊर्जा स्रोतों के विवेकपूर्ण और पर्यावरण-सम्मत उपयोग पर केंद्रित है। सतत संसाधन प्रबंधन के तहत विभिन्न रणनीतियों और नीतियों का विकास किया जाता है ताकि संसाधनों की न्यूनतम बर्बादी हो, पुनर्चक्रण और पुनः उपयोग को प्रोत्साहित किया जाए, और पारिस्थितिकीय संतुलन बनाए रखा जा सके। इसके लिए सरकारों, संगठनों और समुदायों के बीच सहयोग और भागीदारी आवश्यक है।

मुख्य शब्द: सततता (Sustainability), संसाधन प्रबंधन (Resource Management), पुनर्चक्रण (Recycling), पुनः उपयोग (Reuse), प्राकृतिक संसाधन (Natural Resources), पारिस्थितिकीय संतुलन (Ecological Balance), ऊर्जा स्रोत (Energy Sources), जल संरक्षण (Water Conservation),।

प्रारूप

1. परिचय
2. परिभाषा
3. महत्व
4. सतत संसाधन प्रबंधन की आवश्यकता
5. सतत संसाधन प्रबंधन के सिद्धांत

6. सतत संसाधन प्रबंधन के लाभ
7. चुनौतियाँ और समाधान
8. उपसंहार
9. संदर्भ

परिचय

सतत संसाधन प्रबंद्धन (Sustainable Resource Management) का महत्व बढ़ते वैश्विक पर्यावरणीय संकट और सीमित प्राकृतिक संसाधनों के मद्देनजर अत्यधिक बढ़ गया है। यह दृष्टिकोण संसाधनों के न्यायसंगत और विवेकपूर्ण उपयोग को सुनिश्चित करने पर केंद्रित है ताकि आने वाली पीढ़ियाँ भी इनका लाभ उठा सकें। इस अध्याय में, हम सतत संसाधन प्रबंधन के विभिन्न पहलुओं, इसकी आवश्यकता और इसके लाभों पर विस्तार से चर्चा करेंगे।

सतत संसाधन प्रबंधन की परिभाषा

सतत संसाधन प्रबंधन (Sustainable Resource Management) वह प्रक्रिया है जिसके द्वारा प्राकृतिक संसाधनों का उपयोग, संरक्षण और प्रबंधन इस प्रकार किया जाता है कि वे वर्तमान पीढ़ियों की आवश्यकताओं को पूरा करते हुए भविष्य की पीढ़ियों के लिए भी उपलब्ध रहें। इसमें पर्यावरणीय संतुलन बनाए रखने, आर्थिक विकास को बढ़ावा देने और सामाजिक समृद्धि को प्रोत्साहित करने के लिए विभिन्न रणनीतियाँ शामिल हैं। यह अवधारणा पर्यावरणीय संरक्षण, आर्थिक विकास और सामाजिक समृद्धि के बीच संतुलन बनाने पर केंद्रित है।

World Commission on Environment and Development, 1987 के अनुसार- सतत संसाधन प्रबंधन का तात्पर्य उन प्रथाओं और नीतियों से है जो संसाधनों के दीर्घकालिक उपयोग को सुनिश्चित करने के साथ-साथ पर्यावरणीय संतुलन को बनाए रखती हैं, इसमें प्राकृतिक संसाधनों का कुशलतापूर्वक और जिम्मेदाराना उपयोग शामिल है।

महत्व

सतत संसाधन प्रबंधन का मुख्य उद्देश्य पर्यावरणीय क्षरण को रोकना और संसाधनों की उपलब्धता को बनाए रखना है। यह न केवल पर्यावरण संरक्षण के लिए आवश्यक है, बल्कि आर्थिक और सामाजिक स्थिरता के लिए भी महत्वपूर्ण है। उदाहरण के लिए, जल संसाधनों का सतत प्रबंधन कृषि उत्पादन, पेयजल आपूर्ति और जैव विविधता के संरक्षण के लिए अनिवार्य है (Gleick, 2003)।

प्राकृतिक संसाधन जैसे जल, मिट्टी, वायु, खनिज और वन हमारे जीवन के लिए अत्यंत महत्वपूर्ण हैं।

- **जल:** जीवन के लिए सबसे महत्वपूर्ण संसाधन। यह कृषि उत्पादन, पेयजल आपूर्ति उद्योग और जैव विविधता के संरक्षण के लिए अनिवार्य है (Gleick, 2003)। अन्य कई क्षेत्रों में उपयोग होता है।
- **मिट्टी:** कृषि उत्पादन के लिए आवश्यक। यह पौधों को पोषक तत्व प्रदान करती है।
- **वायु:** स्वच्छ वायु स्वास्थ्य के लिए आवश्यक है। प्रदूषित वायु कई स्वास्थ्य समस्याओं का कारण बन सकती है।
- **खनिज:** औद्योगिक और तकनीकी विकास के लिए आवश्यक। ये आर्थिक विकास में महत्वपूर्ण भूमिका निभाते हैं।
- **वन:** वन कई प्रकार की वनस्पतियों और जीवों का घर होते हैं और पर्यावरण संतुलन बनाए रखने में मदद करते हैं।

ये संसाधन न केवल हमारी दैनिक आवश्यकताओं को पूरा करते हैं, बल्कि आर्थिक विकास में भी महत्वपूर्ण भूमिका निभाते हैं। इसलिए, इन संसाधनों का संरक्षण और उनका सतत उपयोग अत्यंत आवश्यक है।

सतत संसाधन प्रबंधन की आवश्यकता

1. **पर्यावरणीय संतुलन बनाए रखना:** अत्यधिक दोहन और प्रदूषण के कारण पर्यावरणीय संतुलन बिगड़ रहा है। यह जलवायु परिवर्तन, जैव विविधता की हानि और पारिस्थितिक तंत्र के विघटन का कारण बन सकता है।सतत संसाधन प्रबंधन के माध्यम से पर्यावरण को संरक्षित किया जा सकता है।
2. **भविष्य की पीढ़ियों के लिए संरक्षण:** संसाधनों का अनियंत्रित उपयोग आने वाली पीढ़ियों के लिए संकट खड़ा कर सकता है। इसलिए, यह जरूरी है कि हम संसाधनों का इस तरह से उपयोग करें कि वे भविष्य में भी उपलब्ध रहें।
3. **आर्थिक स्थिरता:** संसाधनों का सतत उपयोग आर्थिक स्थिरता को बढ़ावा देता है। यह दीर्घकालिक आर्थिक विकास और स्थिरता सुनिश्चित करता है, जिससे सभी लोगों के जीवन स्तर में सुधार होता है।
4. **सामाजिक समानता:** सतत संसाधन प्रबंधन से संसाधनों का समान वितरण सुनिश्चित होता है, जिससे सामाजिक असमानता में कमी आती है और सभी लोगों को उनके अधिकार प्राप्त होते हैं।

सतत संसाधन प्रबंधन के सिद्धांत

1. **पुनर्चक्रण** (Recycling): पुनर्चक्रण से उपयोग किए गए संसाधनों का पुनः उपयोग किया जा सकता है। कचरे को कम करने और पुनर्नवीनीकरण सामग्रियों के उपयोग

को प्रोत्साहित करने के लिए उपाय लागू किए जाते हैं। इससे संसाधनों की खपत में कमी आती है और पर्यावरण पर दबाव कम होता है (Grosse, 2010)द्य उदाहरण के लिए, प्लास्टिक, कागज, धातु और कांच के उत्पादों का पुनर्चक्रण किया जा सकता है।

2. **पुनःप्रयोग** (Reuse): पुनःप्रयोग का अर्थ है वस्तुओं का बार-बार उपयोग करना। इससे संसाधनों की बचत होती है, अपशिष्ट उत्पादन कम होता है और पर्यावरण पर दबाव कम होता है। उदाहरण के लिए, पानी की बोतलें, कपड़े, और इलेक्ट्रॉनिक उपकरणों का पुनःप्रयोग किया जा सकता है।

3. **प्राकृतिक संसाधनों का संरक्षण:** प्राकृतिक संसाधनों का संरक्षण उनके उपयोग को नियंत्रित करके किया जाता है। यह सुनिश्चित करता है कि संसाधन दीर्घकालिक रूप से उपलब्ध रहें। इसमें वन, जल, मिट्टी, और खनिज जैसे प्राकृतिक संसाधनों का संरक्षण शामिल है। वनों की कटाई को रोकने, जल संरक्षण प्रथाओं को अपनाने और मृदा अपरदन को कम करने पर ध्यान केंद्रित किया जाता है (Foley et al., 2005)।

4. **नवीकरणीय ऊर्जा का उपयोग:** नवीकरणीय ऊर्जा स्रोतों जैसे सौर ऊर्जा, पवन ऊर्जा और जैव ऊर्जा का उपयोग संसाधनों के सतत प्रबंधन में सहायक होता है। यह ऊर्जा स्रोत पर्यावरण को कम से कम नुकसान पहुँचाते हैं और भविष्य में ऊर्जा की आवश्यकता को पूरा करने में सहायक होते हैं। ऊर्जा संसाधनों का सतत प्रबंधन भी महत्वपूर्ण है, इसमें नवीकरणीय ऊर्जा स्रोतों का अधिक से अधिक उपयोग बढ़ाना शामिल है (REN21, 2020)।

5. **शिक्षा और जागरूकता:** सतत संसाधन प्रबंधन के महत्व को समझाने और इसे प्रोत्साहित करने के लिए शिक्षा और जागरूकता अभियान चलाए जाते हैं (Tilbury, 2011)। इसके महत्व को समझाने और इसे प्रोत्साहित करने के लिए शिक्षा और जागरूकता अभियान अत्यंत महत्वपूर्ण हैं। कुछ प्रमुख उदाहरण निम्नलिखित हैं:

- **स्वच्छ भारत अभियान** (Clean India Mission): यह अभियान भारत सरकार द्वारा चलाया गया एक प्रमुख अभियान है, जिसका उद्देश्य स्वच्छता, अपशिष्ट प्रबंधन और खुले में शौच को समाप्त करना है। इस अभियान के तहत, लोगों को स्वच्छता के महत्व के बारे में जागरूक किया जाता है और स्वच्छता बनाए रखने के लिए उन्हें प्रशिक्षित किया जाता है। उदाहरण के लिए, इस अभियान के माध्यम से ग्रामीण क्षेत्रों में शौचालय निर्माण को बढ़ावा दिया गया और लोगों को खुले में शौच के दुष्परिणामों के बारे में जानकारी दी गई।
- **पर्यावरण शिक्षा कार्यक्रम** (Environmental Education Program): यह कार्यक्रम विभिन्न देशों में स्कूलों और कॉलेजों में पर्यावरण शिक्षा को बढ़ावा देता है।

इसके अंतर्गत छात्रों को पर्यावरण संरक्षण और सतत संसाधन प्रबंधन के महत्व के बारे में शिक्षित किया जाता है। उदाहरण के लिए, अमेरिका में "Project Learning Tree" एक प्रमुख पर्यावरण शिक्षा कार्यक्रम है जो छात्रों को पर्यावरणीय मुद्दों पर जागरूक करता है। इसके तहत कक्षाओं में पर्यावरणीय गतिविधियों का आयोजन किया जाता है और छात्रों को प्रकृति के संरक्षण के महत्व के बारे में बताया जाता है।

- **जल सुरक्षा अभियान** (Water Security Campaign): यह अभियान जल संरक्षण और सतत जल प्रबंधन को बढ़ावा देने के लिए चलाया जाता है। भारत में, "जल शक्ति अभियान" एक प्रमुख पहल है जो जल संरक्षण के महत्व पर जोर देती है और समुदायों को जल स्रोतों के संरक्षण के लिए प्रोत्साहित करती है। इस अभियान के तहत, जल स्रोतों की सफाई, वर्षा जल संचयन और जल संरक्षण के अन्य उपायों को प्रोत्साहित किया जाता है।
- **सतत जीवनशैली के लिए शिक्षा** (Education for Sustainable Living): संयुक्त राष्ट्र का "UNESCO Global Action Programme (GAP) on Education for Sustainable Development" एक प्रमुख पहल है जो सतत विकास लक्ष्यों (SDGs) को प्राप्त करने के लिए शिक्षा को बढ़ावा देता है। यह कार्यक्रम शिक्षा के माध्यम से लोगों को सतत जीवनशैली अपनाने के लिए प्रेरित करता है। उदाहरण के लिए, इस कार्यक्रम के तहत विभिन्न कार्यशालाओं, सेमिनारों और प्रशिक्षण सत्रों का आयोजन किया जाता है ताकि लोग पर्यावरणीय स्थिरता के महत्व को समझ सकें और अपने जीवन में इसे लागू कर सकें।
- **ग्रीन स्कूल्स प्रोग्राम** (Green Schools Program): यह कार्यक्रम स्कूलों में पर्यावरणीय जागरूकता और सतत प्रथाओं को बढ़ावा देने के लिए चलाया जाता है। भारत में "Centre for Science and Environment (CSE)" का ग्रीन स्कूल्स प्रोग्राम एक प्रमुख पहल है जो स्कूलों को पर्यावरणीय मुद्दों पर शिक्षित करता है और उन्हें हरित प्रथाओं को अपनाने के लिए प्रेरित करता है। उदाहरण के लिए, इस कार्यक्रम के तहत स्कूलों को ऊर्जा की बचत, जल संरक्षण, और अपशिष्ट प्रबंधन के लिए प्रेरित किया जाता है और उनके प्रयासों को प्रमाणित किया जाता है।

इन अभियानों के माध्यम से शिक्षा और जागरूकता को बढ़ावा देकर सतत संसाधन प्रबंधन को प्रभावी रूप से लागू किया जा सकता है। इन कार्यक्रमों के तहत न केवल पर्यावरणीय ज्ञान का प्रसार किया जाता है, बल्कि समुदायों को सतत प्रथाओं को अपनाने के लिए भी प्रेरित किया जाता है

सतत संसाधन प्रबंधन के लाभ

1. **पर्यावरण संरक्षण:** सतत संसाधन प्रबंधन से पर्यावरण को होने वाले नुकसान में कमी आती है। यह वायु, जल, और मिट्टी के प्रदूषण को कम करता है और जैव विविध ता को संरक्षित करता है।
2. **आर्थिक लाभ:** संसाधनों के दक्ष उपयोग से उत्पादन लागत में कमी आती है। यह दीर्घकालिक आर्थिक विकास और स्थिरता सुनिश्चित करता है। साथ ही, नए रोजगार के अवसर भी उत्पन्न होते हैं।
3. **सामाजिक समृद्धि:** सतत संसाधन प्रबंधन से समाज में संसाधनों का समान वितरण होता है। यह सामाजिक असमानता को कम करता है और समाज में समृद्धि लाता है।
4. **स्वास्थ्य लाभ:** स्वच्छ और हरित पर्यावरण से मानव स्वास्थ्य में सुधार होता है। प्रदूषण कम होने से विभिन्न बीमारियों का खतरा कम होता है और जीवन स्तर में सुधार होता है।

सतत संसाधन प्रबंधन के चुनौतियाँ और समाधान

सतत संसाधन प्रबंधन के समक्ष कई चुनौतियाँ हैं, जैसे कि नीति निर्माण में कमी, आर्थिक अवरोध, और जागरूकता की कमी। इन चुनौतियों से निपटने के लिए कुछ उपाय सुझाए गए हैं:

1. **शिक्षा और जागरूकता:** सतत संसाधन प्रबंधन के लिए शिक्षा और जागरूकता अत्यंत महत्वपूर्ण है। लोगों को संसाधनों के महत्व और उनके सतत उपयोग के तरीकों के बारे में शिक्षित किया जाना चाहिए। स्कूलों और समुदायों में सतत संसाधन प्रबंधन पर शिक्षा और प्रशिक्षण कार्यक्रम चलाए जाने चाहिए (UNESCO] 2014)।
2. **नीतियाँ और कानून:** सरकारों को सतत संसाधन प्रबंधन के लिए उचित नीतियाँ और कानून बनाने चाहिए। यह सुनिश्चित करेगा कि संसाधनों का उपयोग नियंत्रित और न्यायसंगत हो। सरकारों को सख्त और स्पष्ट नीतियाँ बनानी चाहिए जो सतत संसाधन प्रबंधन को बढ़ावा दें (Meadows et al.,2004)।
3. **स्थानीय समुदाय की भागीदारी:** स्थानीय समुदायों की भागीदारी से सतत संसाधन प्रबंधन को प्रभावी ढंग से लागू किया जा सकता है। सामुदायिक प्रयासों से संसाधनों का संरक्षण और पुनःप्रयोग संभव है।
4. **प्रौद्योगिकी का उपयोग:** संसाधनों के सतत प्रबंधन के लिए नवीनतम प्रौद्योगिकी का उपयोग किया जा सकता है। यह न केवल संसाधनों के उपयोग को दक्ष बनाता है, बल्कि पर्यावरण पर होने वाले प्रभाव को भी कम करता है।
5. **अर्थिक प्रोत्साहन:** सतत प्रथाओं को अपनाने के लिए आर्थिक प्रोत्साहन और सब्सिडी प्रदान की जानी चाहिए (Costanza et al.,1997)।

उपसंहार

सतत संसाधन प्रबंधन वर्तमान समय की महत्वपूर्ण आवश्यकता है। इसके माध्यम से न केवल वर्तमान पीढ़ी की आवश्यकताओं को पूरा किया जा सकता है, बल्कि भविष्य की पीढ़ियों के लिए भी संसाधनों को सुरक्षित रखा जा सकता है। यह पर्यावरण, आर्थिक और सामाजिक दृष्टिकोण से अत्यंत महत्वपूर्ण है। हमें अपने संसाधनों का उपयोग सतर्कता और संवेदनशीलता के साथ करना चाहिए ताकि वे हमारे और आने वाली पीढ़ियों के लिए उपलब्ध रहें। सतत संसाधन प्रबंधन को सफलतापूर्वक लागू करने के लिए समन्वित प्रयासों और ठोस नीतियों की आवश्यकता है। सतत संसाधन प्रबंधन का अनुसरण करके ही हम एक समृद्ध, स्वस्थ और संतुलित भविष्य की ओर बढ़ सकते।

संदर्भ

Costanza, R., d'Arge, R., de Groot, R., Farber, S., Grasso, M., Hannon, B., ... & van den Belt, M. (1997). The value of the world's ecosystem services and natural capital. *Nature*, 387(6630), 253-260.

Foley, J. A., De Fries, R., Asner, G. P., Barford, C., Bonan, G., Carpenter, S. R., ... & Snyder, P. K. (2005). Global consequences of land use. Science, 309(5734), 570-574.

Gleick, P. H. (2003). Water use. Annual Review of Environment and Resources, 28(1), 275-314.

Grosse, F. (2010). Is recycling "part of the solution"? The role of recycling in an expanding society and a world of finite resources. Sustainability, 2(4), 1403-1426.

Meadows, D. H., Meadows, D. L., Randers, J., & Behrens III, W. W. (2004). The limits to growth: The 30-year update. Chelsea Green Publishing.

REN21. (2020). Renewables 2020 Global Status Report. REN21 Secretariat.

Tilbury, D. (2011). Education for sustainable development: An expert review of processes and learning. UNESCO.

UNESCO. (2014). Education for Sustainable Development. United Nations Educational, Scientific and Cultural Organization.

World Commission on Environment and Development. (1987). Our common future. Oxford University Press.

अध्याय 25

पर्यावरण शिक्षा एवं जागरूकता

डॉ. कावेरी दाभड़कर[1], डॉ. फूहार मेघावाले[2]

[1]सहायक प्राध्यापक (भूगोल), शास. बिलासा कन्या स्नातकोर महाविद्यालय, बिलासपुर (छ.ग.)

[2]शोधार्थी, मैट्स विश्वविद्यालय, रायपुर (छ.ग.)

पर्यावरण, सृष्टि के मूलभूत तत्वों "क्षिति, जल, पावक, गगन, समीरा" से निर्मित मानव की अस्मिता का मूल है। पृथ्वी पर जीवन के लिए उत्तरदायी है। पृथ्वी, जो अगणित संसाधनों से युक्त है, हमेशा मनुष्य जाति के लिए निर्वहन का साधन है। पृथ्वी से ही वनस्पति, अन्न, भोजन, वस्त्र एवं मकान जैसी आवश्यकता की आपूर्ति होती है। प्रकृति के गोद में जन्मा मनुष्य हमेशा ही प्रकृति की क्षमताओं का उपभोक्ता रहा है। 21 वीं सदी के पूर्वार्द्ध में पृथ्वी तल के संसाधनों जैसे जल, वायु, खनिज, वनस्पति, मिट्टी इत्यादि एवं मानव द्वारा इसके उपभोग के मध्य संतुलन स्थापित था। बढ़ती जनसंख्या वृद्धि, मानवीय बुद्धि की तीव्रता एवं बढ़ते औद्योगिक विकास ने संसाधनों एवं उपयोग के बीच असंतुलन की स्थिति उत्पन्न हुई। परिण ामस्वरूप बढ़ती आवश्यकताओं की पूर्ति हेतु मानव पृथ्वी का अंधाधुंध करने लगा, जिसकी परिणिती 'पर्यावरण अवनयन'। वृक्षों की अंधाधुंध कटाई, पर्यावरणीय चक्र की सामान्य गति में परिवर्तन, जलवायु परिवर्तन, न्यून उत्पादकता, मिट्टी की अनुर्वरतः प्रदूषण (जल, वायु, ध्वनि) जैसी गंभीर मुद्दे यक्ष प्रश्न बन गए हैं। बढ़ते हुए पर्यावरणीय संकट का सामना वैश्विक स्तर पर सभी राष्ट्रों द्वारा किया जा रहा है। संकट के समाधान हेतु वैश्विक हल भी ढूंढे जा रहे हैं। पर्यावरणीय सम्मेलनों के माध्यम से प्रकृति संरक्षण के समय-समय पर प्रयत्न होते रहे है।

तथापि वैश्विक मंच पर पर्यावरण संरक्षण के परिप्रेक्ष्य में पर्यावरणीय शिक्षा एवं पर्यावरणीय जागरूकता को अनिवार्य घटक माना जा रहा है। पर्यावरणीय शिक्षा मानवीय व्यवहार के उन पहलुओं से संबंधित है जो जैविक, भौतिक पर्यावरण एवं मानव के अंर्तसंबंधों से प्रत्यक्ष रूप से प्रभावित है, साथ ही यह प्रकृति के साथ मानव के अंर्तप्रक्रिया के प्रति समझ को इंगित करती है।

आज पर्यावरण में व्याप्त प्रदूषण के निवारण का सबसे कारगर उपाय मानव समुदाय को पर्यावरण के प्रति सचेत करना। मनुष्य के द्वारा पृथ्वी के अतिदोहन एवं चहुंव्यापी दोहन के कारण ही प्रदूषण एवं पर्यावरणीय समस्याएं उद्भवित हुई है। यदि समय रहते हम पर्यावरण के प्रति अपने उत्तरदायित्व को निभाएंगे नहीं तो समूची मानव सभ्यता खतरे में पड़ जावेगी। अतः पर्यावरणीय शिक्षा की आवश्यता न केवल समसामयिक है अपितु पृथ्वी पर जीवन के लिए उत्तरदायी है। यूनेस्कों ने पर्यावरणीय शिक्षा की आवश्यकता व परिभाषा के संबंध में लिखा है –

पर्यावरणीय शिक्षा, पर्यावरण संरक्षण के उद्देश्यों/लक्ष्यों को साधने का उपागम है, तरीका है। यह किसी विज्ञान शाख की विषयवस्तु नहीं अपितु आजीवन चलने वाला क्षेत्रीय अध्ययन है।

वस्तुतः पर्यावरणीय शिक्षा समूचे मानवता के जीवन गुणवत्ता के संवर्धन एवं विकास का एक सशक्त साधन है, जो कि पर्यावरण रक्षा और पर्यावरण संवर्धन के परिप्रेक्ष्य में आवश्यक है।

पर्यावरण शिक्षा एक ऐसी शैक्षणिक प्रक्रिया है जिसके अंतर्गत व्यक्ति में पर्यावरण जागरूकता के प्रति सकारात्मक ज्ञान का विकास होता है। परिणामस्वरूप व्यक्ति पर्यावरण के संरक्षण, उपभोग एवं सुरक्षा के प्रति संवेदनशील हो उसका उपभोग भविष्य की संभावनाओं के परिप्रेक्ष्य में सम्यक रूप से करने की पहल करता है। पर्यावरण शिक्षा के कारण ही मनुष्य प्राकृतिक संसाधनों – जल, वायु, मिट्टी के उपयोग में सजग रहकर पर्यावरण की गुणवत्ता के पोषण में महत्वपूर्ण भूमिका का निर्वहन करता है, जिससे अंततः मानव जीवन की गुणवत्ता में संवर्धन होता है।

पर्यावरण शिक्षा की आवष्यकता

मानव जीवन की निरंतरता प्राकृतिक संसाधनों यथा ऊर्जा, जल, खनिज, वन इत्यादि पर निर्भर है। मनुष्य अपने ज्ञान एवं तकनीकी के प्रयोग से इन संसाधनों को परिमार्जित कर वस्तुओं, मशीन इत्यादि जीवनपयोगी साधनों का निर्माण करता है जो मानव जीवन की, स्वास्थ्य की, गुणवत्ता की प्रथम आवश्यकता है। वस्तुतः समस्त मानव जीवन का मूल आधार ही है ये प्राकृतिक संसाधन। अतः यह आवश्यक है कि सीमित मात्रा में उपलब्ध इस संसाधनों का हम सम्यक उपभोग करें अन्यथा पर्यावरणीय अवनयन, प्रदूषण एवं पारिस्थितिकीय असंतुलन जनित अनेक समस्याएँ उद्भवित हो रही है, जो आज एक खतरा बन गयी है। विकास की अंधी दौड़ में हमने अपने पर्यावरण का निरंकुश उपयोग किया है। परिणामतः जलवायु परिवर्तन, अम्लीय वर्षा, सूखा, बाढ़ जैसी समस्याएँ आज समूचे भारत को भयाक्रांत कर रही है। अतः समय रहते यदि हम पर्यावरण के प्रति जागरूक न होंगे तो समूचे मानव जाति के विनाश का कारण होगा। पर्यावरण शिक्षा का औचित्य इसलिए है कि आज की जागरूता एवं ज्ञान भविष्य की सुरक्षा की गारंटी है। अस्तु पर्यावरणीय शिक्षा अभीष्ट है।

पर्यावरणीय शिक्षा एवं संयुक्त राष्ट्र संघ

संयुक्त राष्ट्र संघ विभिन्न प्रयासों के माध्यम से हमेशा से पर्यावरणीय सुरक्षा, शाश्वत विकास, संसाधन संरक्षण जैसे मुद्दों पर संवेदनशील रहा है।

1. मान्ट्रियल समझौता
2. टोक्यो सम्मेलन
3. रियो डि जेनेरो सम्मेलन
4. पृथ्वी सम्मेलन

यूनेस्को के पर्यावरणीय संरक्षण (1989) के अंतर्गत पर्यावरणीय शिक्षा की आवश्यकता एवं उद्देश्य को रेखांकित किया गया है :-

जागरूकता

पर्यावरणीय शिक्षा का प्रमुख उद्देश्य संपूर्ण पर्यावरणीय समस्याओं के प्रति लोगों को जागरूक करना जिसके फलस्वरूप पर्यावरणीय समस्याओं की पहचान कर उसका स्थानापन्न् खोज सकें।

ज्ञान

पर्यावरणीय शिक्षा व्यक्ति अथवा सामाजिक समूहों की स्थानीय पर्यावरणीय को पहचानने की क्षमता में वृद्धि करती है, जिससे वे आसपास के पर्यावरण के अनुरूप परिवर्तन की चेष्टा कर सकते हैं।

व्यवहार

पर्यावरणीय शिक्षा व्यक्तियों या सामाजिक समूहों को सामाजिक मूल्यों के प्रति जागरूक बनाती है, जो उनमें पर्यावरणीय सुरक्षा संबंधी कार्यां में सक्रिय सहभागिता के प्रति संवेदनशीलता विकसित करती है।

कौशल

पर्यावरणीय शिक्षा का उद्देश्य मनुष्य में कौशल विकास करना है, जिससे वे अपने आसपास के पर्यावरणीय समस्याओं की पहचान कर उसके समाधान की दिशा में प्रयत्न कर सकें।

सहभागिता

पर्यावरणीय शिक्षा व्यक्तियों में पर्यावरण के सुरक्षा के प्रति लोगों में उत्तरदायित्व का बोध जागृत करती है, जिससे पृथ्वी के समस्त मानव पर्यावरणीय समस्याओं के निदान में सक्रिय सहभागी बने। पर्यावरणीय शिक्षा के प्रति उपरोक्त उद्देश्यों की पूर्णता की दिशा में अनेक वैश्विक कार्यक्रमों यूनेस्कों द्वारा संचालित किये जा रहे हैं।

भारत में पर्यावरणीय शिक्षा एवं जागरूकता अभियान

भारत में भी पर्यावरणीय शिक्षा को संविधानसम्मत सुरक्षा के साथ पाठ्यक्रमों का अभिन्न भाग बनाया गया है। औपचारिक शैक्षणिक पाठ्यक्रम के अतिरिक्त अनेक पर्यावरणीय सुरक्षा के उपाय भारत सरकार द्वारा किए गए हैं, यथा :-

1. पर्यावरण सुरक्षा कानून
2. पर्यावरणीय सुरक्षा अधिनियम
3. जल प्रदूषण, नियंत्रण अधिनियम 1974
4. वन्य जीव संरक्षण अधिनियम 1986, 1972
5. वायु प्रदूषण नियंत्रण अधिनियम 1981
6. पर्यावरण संरक्षण कानून 1986

इसके अतिरिक्त अनेक पर्यावरण जागरूकता अभियानों के माध्यम से पर्यावरणीय शिक्षा को जन-जन तक पहुंचाया जा रहा है। भारत सरकार के पर्यावरण वन एवं जलवायु परिवर्तन मंत्रालय द्वारा छंजपवदंस ळतममद ब्वतचे (NGC) वर्ष 2001-2002 से क्रियान्वित किया जा रहा है, जिसके अंतर्गत शालाओं में म्बव ब्सनइ का गठन कर विभिन्न पर्यावरणीय कार्यविधियों को करना, जिससे देश का युवा, देश के भविष्य को आज से ही पर्यावरण के प्रति सजग करें। यह गतिविधि 2021-2022 में अनवरत् रही।

भारत में 2019 में केंद्रीय विद्यालयों में भी इको क्लब के गठन हेतु आदेश किया गया। शिक्षा मंत्रालय ने भी इस कार्य में सहभागिता दी एवं देश व राज्य के सभी शैक्षणिक संस्थाओं में इको-क्लब के गठन को अनिवार्य कर दिया। नई एवं राष्ट्रीय शिक्षा नीति 2020 के अंतर्गत देश के सभी राज्यों के शालाओं एवं उच्च शिक्षा संस्थानों में इको क्ल्ब एवं इसकी गतिविधियों को प्रमुखता दी गई है।

राष्ट्रीय स्तर पर पर्यावरणीय शिक्षा एवं जागरूकता को संर्वधित करने के उद्देश्य से वर्ष 1986 से राष्ट्रीय पर्यावरण जागरूकता अभियान का सूत्रपात हुआ, जिसके अंतर्गत देश के सभी महाविद्यालयों, शालाओं, गैर सरकारी संगठनों, विश्वविद्यालयों में अल्प विळाय सहायता के माध्यम से पर्यावरण जागरूकता एवं शिक्षा संबंधी अनेक कार्यक्रम आयोजित किये जाते हैं यथा- संगोष्ठी, कार्यशालाएं, चर्चा, पोस्टर प्रतियोगिता, रैली, प्रदर्शनी के साथ-साथ वृक्षारोपण, घरेलू अपशिष्ट प्रबंधन कार्यशालाएं भी आयोजित किये जाते हैं।

भारत में पर्यावरणीय शिक्षा एवं जागरूकता के औपचारिक एवं अनौपचारिक प्रयत्न किए गए हैं।

औपचारिक संदर्भ

पर्यावरणीय शिक्षा शालेय, महाविद्यालय एवं विश्वविद्यालय परिसरों में प्रदत्त प्रशिक्षण, अध्ययन, अध्यापन इसके अंतर्गत शामिल है। विद्यार्थियों को पाठयक्रम के माध्यम से पर्यावरणीय शिक्षा के साथ-साथ पर्यावरण के प्रति सजग रहने हुत अध्यापन किया जाता है। पर्यावरणीय अध्ययन में डिग्रियाँ, शोध, प्रोजेक्ट इत्यादि क्रियान्वित किये जा रहे हैं।

अनौपचारिक संदर्भ

हमारे राष्ट्र में अनेकों युवा शालेय, महाविद्यालय, जमीनी शिक्षा से वंचित है। अतः ऐसे युवाओं के लिए पर्यावरणीय शिक्षा एवं जागरूकता के कार्यक्रम प्रशिक्षण आयोजित करना आवश्यक है। वस्तुतः पर्यावरणीय शिक्षा को अध्ययन कक्षों से बाहर पर्यावरणीय समस्या के प्रति सजग एवं निवारण हेतु सामाजिक समूहों को साथ ले लेकर करने का प्रयास है।

गैर पारंपरिक पर्यावरणीय शिक्षा में विभिन्न आयोजन यथा पाठ्येत्तर गतिविधियाँ, पोस्टर, निबंध प्रतियोगिता, प्रदर्शनी सेमीनार, प्राकृतिक शिविर, प्राकृतिक क्लब की गतिविधियाँ, इत्यादि शामिल है। पारिस्थितिकी विकास क्लब का मूलभूत उद्देश्य ही जनता को पर्यावरणीय, पारिस्थितिकीय समस्याओं के प्रति जागरूक करना है। एनएसएस/एन.सी.सी. की छात्राओं द्वारा गंगा सफाई मिशन, नदी स्वच्छता, समुद्र किनारों की स्वच्छता जैसे कार्यक्रम भी पर्यावरण जागरूकता में वृद्धि करते हैं। नुक्कड़ नाटक, भाषण इत्यादि माध्यमों से भी पर्यावरणीय जागरूकता शिक्षा दी जाती है। पर्यावरणीय शिक्षा के साथ-साथ पर्यावरण जागरूकता भी पर्यावरण के अवनयन को नियंत्रित करने का अचूक उपाय है। विशेषतः पर्यावरण प्रदूषण के परिप्रेक्ष्य में जनजागरूकता अनिवार्य है। मानवीय गतिविधियों ही मुख्यतः पर्यावरणीय प्रदूषण, संसाधनों के अविवेकपूर्ण उपयोग हेतु उत्तरदायी है। अतः प्रत्येक व्यक्ति को पर्यावरण के प्रति जागरूक होना होगा। भारतीय संदर्भों में शासन द्वारा अनेक पर्यावरणीय जागरूकता कार्यक्रम क्रियान्वित किये जाते हैं। तथापि इस दिशा में और भी प्रयत्न किए जाने चाहिए यथा -

1. पर्यावरण संवर्धन कार्यक्रमों में जनता को सहभागिता।
2. प्रचार माध्यमों से पर्यावरणीय जागरूकता का प्रसार।
3. पर्यावरण संबंधी अधिनियमों, विनियमों का निर्धारण।
4. शालेय, महाविद्यालय, विश्वविद्यालय स्तर पर जनजागरूकता एवं विद्यार्थियों की सहभागिता से रैली, ग्राम-गाम जाकर पर्यावरण जागरूकता का प्रयास।
5. जनजागरूकता पर लघु फिल्में, नुक्कड़ नाटक इत्यादि।
6. सामान्यजन के लिये पर्यावरण शिक्षा व जागरूकता पर प्रशिक्षण कार्यक्रम।
7. रेडियो, टीवी के माध्यम से पर्यावरण जागरूकता।

8. वृक्षारोपण, श्रमदान, नदी संरक्षण जैसे पर्यावरणीय मुद्दे पर जनसहयोग।
9. घर-घर जाकर पर्यावरण संरक्षण की छोटी-छोटी बातों को जन सामान्य को जागरूक करना।
10. सामाजिक संगठनों की भूमिका भी पर्यावरण शिक्षा एवं जागरूकता में महत्वपूर्ण है। इन संगठनों द्वारा भी स्थान-स्थान पर जागरूकता शिविरों के माध्यम से पर्यावरण सुरक्षा के प्रति लोगों को उत्तरदायी बनाना चाहिए।
11. शासन को पर्यावरणीय अधिनियमों के कड़ाई से पालन करने के प्रति कटिबद्ध होना होगा।
12. व्यक्तिगत स्तर पर भी पर्यावरण सुरक्षा व प्रदूषण को कम करने प्रयत्न किये जाने चाहिए, यथा

- वृक्ष लगाना
- जल के अपव्यय को रोकना
- ऊर्जा संरक्षण के प्रति सचेत रहना
- वन आवरण को बचाना
- अपशिष्टों के फैलान पर नियंत्रण
- स्वास्थ्य एवं हाइजिन पर नियंत्रण इत्यादि

समग्रतः मानवजीवन के अस्तित्व एवं संसाधनों को अक्षुण्ण रखने हेतु उनका शाश्वत विकास आवश्यक है। समाज को पर्यावरणीय शिक्षा एवं पर्यावरण के प्रति जागरूक करेंगे तो पर्यावरणीय समस्याओं का निदान संभव होगा। अतः पर्यावरण जागरूकता एक जन आंदोलन की भांति भारत के प्रत्येक नागरिक का कर्त्तव्य होना चाहिए।

संदर्भ

अहलूवालिया, एस.पी. एवं बैस, एच.एस. (1992): पर्यावरणीय शिक्षा, आशीष पब्लिशिंग हाउस, नई देहली

नाथ भास्कर (2009): Environment Education and Awareness, Vol- 1, Encyclopedia of life support system.

Barti Dutta (2023): Environmental Education and Public Awareness, "Kachhioli": A Journal of Huminities and social sicence Vol III, issue 1, June 2023.

Elegba Mohmed (2006): Environmental Education and Public Awarness, Paper in workshop on Post Conflict National Plan EVS Managment suddan 18-20 July 2006.

INDEX